# AS-Level Biology

## The Revision Guide

### Exam Board: AQA

Published by Coordination Group Publications Ltd.

From original material by Richard Parsons.

*Editors:*
Ellen Bowness, Joe Brazier, Charlotte Burrows, Tom Cain, Katherine Craig, Andy Park, Laurence Stamford, Jane Towle.

*Contributors:*
Gloria Barnett, Jessica Egan, James Foster, Barbara Green, Brigitte Hurwitt, Liz Masters, Stephen Phillips, Claire Ruthven, Adrian Schmit, Anna-fe Williamson.

*Proofreaders:*
Glenn Rogers, Sue Hocking.

ISBN: 978 1 84762 118 4

*With thanks to Jan Greenway for the copyright research.*

*Data used to construct the graph of BMI and cancer on page 5 reproduced with kind permission from Gillian K Reeves, Kristin Pirie, Valerie Beral et al. Copyright © 2007, BMJ Publishing Group Ltd.*

*MMR graph on page 10 adapted from H. Honda, Y. Shimizu, M. Rutter. No effect of MMR withdrawal on the incidence of autism: a total population study. Journal of Child Psychology and Psychiatry 2005; 46(6):572-579.*

*Data used to construct Herceptin graph on page 10 from M.J. Piccart-Gebhart, et al. Trastuzumab after Adjuvant Chemotherapy in HER2-positive Breast Cancer. NEJM 2005; 353: 1659-72.*

*Data used to construct the Hib graph on page 11 reproduced with kind permission from the Health Protection Agency.*

*With thanks to Cancer Research UK for permission to reproduce the graphs on page 38.*
*Cancer Research UK, http://info.cancerresearchuk.org/cancerstats/types/lung/mortality/, January 2008.*
*Cancer Research UK, http://info.cancerresearchuk.org/cancerstats/types/lung/smoking/, January 2008.*

*Data used to construct asthma and sulfur dioxide graphs on page 39. Source: National Statistics website: www.statistics.gov.uk Crown copyright material is reproduced with the permission of the Controller Office of Public Sector Information (OPSI).*

*Exam question graph on page 39: The Relationship Between Smoke And Sulphur Dioxide Pollution And Deaths During The Great London Smog, December 1952, Source: Wilkins, 1954.*

*Data used to construct the graph on page 45 from R. Doll, R. Peto, J. Boreham, I Sutherland. Mortality in relation to smoking: 50 years' observations on male British doctors. BMJ 2004; 328:1519.*

*Data used to construct the graph on page 84 from National Statistics online. Reproduced under the terms of the Click-Use licence.*

*Graph of skylark population on page 88 from BTO/JNCC Breeding Birds of the Wider Countryside.*

*Graph of rainforest diversity on page 89 from Schulze et al. Biodiversity Indicator Groups of Tropical Land-Use Systems: Comparing Plants, Birds and Insects. Ecological Applications 2004; 14(5) Ecological Society of America.*

*Exam question graph on page 89 from Defra. Reproduced under the terms of the Click-Use licence.*

Groovy website: www.cgpbooks.co.uk
Jolly bits of clipart from CorelDRAW®
Printed by Elanders Hindson Ltd, Newcastle upon Tyne.

# Contents

# The Scientific Process

*'How Science Works' is all about the scientific process — how we develop and test scientific ideas. It's what scientists do all day, every day (well, except at coffee time — never come between a scientist and their coffee).*

## Scientists Come Up with **Theories** — Then **Test Them...**

Science tries to explain **how** and **why** things happen — it **answers questions**. It's all about seeking and gaining **knowledge** about the world around us. Scientists do this by **asking** questions and **suggesting** answers and then **testing** them, to see if they're correct — this is the **scientific process**.

1) **Ask** a question — make an **observation** and ask **why or how** it happens. E.g. why is trypsin (an enzyme) found in the small intestine but not in the stomach?
2) **Suggest** an answer, or part of an answer, by forming a **theory** (a possible **explanation** of the observations) e.g. pH affects the activity of enzymes. (Scientists also sometimes form a **model** too — a **simplified picture** of what's physically going on.)
3) Make a **prediction** or **hypothesis** — a **specific testable statement**, based on the theory, about what will happen in a test situation. E.g. trypsin will be active at pH 8 (the pH of the small intestine) but inactive at pH 2 (the pH of the stomach).
4) Carry out a **test** — to provide **evidence** that will support the prediction (or help to disprove it). E.g. measure the rate of reaction of trypsin at various pH levels.

The evidence supported Quentin's Theory of Flammable Burps.

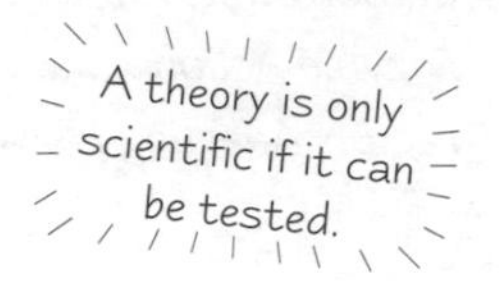


## ...Then They **Tell** Everyone About Their **Results...**

The results are **published** — scientists need to let others know about their work. Scientists publish their results in **scientific journals**. These are just like normal magazines, only they contain **scientific reports** (called papers) instead of the latest celebrity gossip.

1) Scientific reports are similar to the **lab write-ups** you do in school. And just as a lab write-up is **reviewed** (marked) by your teacher, reports in scientific journals undergo **peer review** before they're published.
2) The report is sent out to **peers** — other scientists who are experts in the **same area**. They examine the data and results, and if they think that the conclusion is reasonable it's **published**. This makes sure that work published in scientific journals is of a **good standard**.
3) But peer review **can't guarantee** the science is **correct** — other scientists still need to **reproduce** it.
4) Sometimes **mistakes** are made and flawed work is published. Peer review **isn't perfect** but it's probably the best way for scientists to self-regulate their work and to publish **quality reports**.

## ...Then **Other Scientists** Will **Test** the Theory Too

Other scientists read the published theories and results, and try to **test the theory** themselves. This involves:

- Repeating the **exact same experiments**.
- Using the theory to make **new predictions** and then testing them with **new experiments**.

## If the **Evidence** Supports a Theory, It's **Accepted — for Now**

1) If all the experiments in all the world provide good evidence to back it up, the theory is thought of as **scientific 'fact'** (for now).
2) But it will never become **totally indisputable** fact. Scientific **breakthroughs or advances** could provide new ways to question and test the theory, which could lead to **new evidence** that **conflicts** with the current evidence. Then the testing starts all over again...

And this, my friend, is the **tentative nature of scientific knowledge** — it's always **changing** and **evolving**.

# The Scientific Process

*So scientists need evidence to back up their theories. They get it by carrying out experiments, and when that's not possible they carry out studies. But why bother with science at all? We want to know as much as possible so we can use it to try and improve our lives (and because we're nosy).*

## Evidence Comes from Lab Experiments...

1) Results from **controlled experiments** in **laboratories** are **great**.
2) A lab is the easiest place to **control variables** so that they're all **kept constant** (except for the one you're investigating).
3) This means you can draw meaningful **conclusions**.

For example, if you're investigating how temperature affects the rate of an enzyme-controlled reaction you need to keep everything but the temperature constant, e.g. the pH of the solution, the concentration of the solution etc.

## ...and Well-Designed Studies

1) There are things you **can't** investigate in a lab, e.g. whether stress causes heart attacks. You have to do a study instead.
2) You still need to try and make the study as controlled as possible to make it **more reliable**. But in reality it's **very hard** to control **all the variables** that **might** be having an effect.
3) You can do things to help, e.g. have **matched groups** — **choose two groups** of people (those who have quite stressful jobs and those who don't) who are **as similar as possible** (same mix of ages, same mix of diets etc.). But you can't easily rule out every possibility.

Samantha thought her study was very well designed — especially the fitted bookshelf.

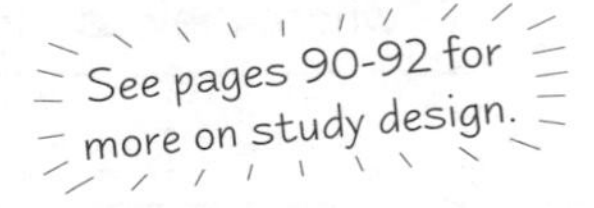


## Society Makes Decisions Based on Scientific Evidence

1) Lots of scientific work eventually leads to **important discoveries** or breakthroughs that could **benefit humankind**.
2) These results are **used by society** (that's you, me and everyone else) to **make decisions** — about the way we live, what we eat, what we drive, etc.
3) All sections of society use scientific evidence to make decisions, e.g. politicians use it to devise policies and individuals use science to make decisions about their own lives.

Other factors can **influence** decisions about science or the way science is used:

**Economic factors**

- Society has to consider the **cost** of implementing changes based on scientific conclusions — e.g. the **NHS** can't afford the most expensive drugs without **sacrificing** something else.
- Scientific research is **expensive** so companies won't always develop new ideas — e.g. developing new drugs is costly, so pharmaceutical companies often only invest in drugs that are likely to make them **money**.

**Social factors**

- **Decisions** affect **people's lives** — E.g. scientists may suggest **banning smoking** and **alcohol** to prevent health problems, but shouldn't **we** be able to **choose** whether **we** want to smoke and drink or not?

**Environmental factors**

- Scientists believe **unexplored regions** like remote parts of rainforests might contain **untapped drug** resources. But some people think we shouldn't **exploit** these regions because any interesting finds may lead to **deforestation** and **reduced biodiversity** in these areas.

## So there you have it — how science works...

*Hopefully these pages have given you a nice intro to how science works, e.g. what scientists do to provide you with 'facts'. You need to understand this, as you're expected to know how science works — for the exam and for life.*

# Disease

*What a lovely way to start a book... two pages on disease. Nice.*

## Pathogens Cause Infectious Diseases

1) A pathogen is any **organism** that **causes disease.**
2) Pathogens include **microorganisms** and some larger organisms, such as **tapeworms.**
3) Pathogenic microorganisms include some **bacteria**, some **fungi** and all **viruses.**

## Pathogens can Penetrate an Organism's Interface with the Environment

Pathogens need to **enter** the body to cause disease — they **get in** through an organism's **surface of contact** (**interface**) with the **environment**, e.g. nose, eyes, a cut. An organism has **three** main interfaces with the environment — the **gas-exchange system**, the **skin** and the **digestive system**.

Gas-Exchange System — If you breathe in **air** that contains **pathogens**, most of them will be trapped in **mucus** lining the lung epithelium (the outer layer of cells in the passages to the lungs). These cells also have **cilia** (hair-like structures) that **beat** and **move** the mucus up the trachea to the mouth, where it's removed. Unfortunately, some pathogens are still able to reach the **alveoli** where they can **invade** cells and cause **damage.**

Skin — If you **damage** your skin, **pathogens** on the surface can enter your **bloodstream**. The blood **clots** at the area of damage to **prevent** pathogens from entering, but some may get in **before** the clot forms.

Digestive System — If you **eat** or **drink food** that contains **pathogens**, most of them will be **killed** by the **acidic** conditions of the **stomach**. However, some may **survive** and pass into the intestines where they can invade **cells** of the **gut wall** and cause disease.

## Pathogens Cause Disease by Producing Toxins and Damaging Cells

Despite our **protective mechanisms**, pathogens can still **successfully enter** our bodies. Once inside, they **cause disease** in **two** main ways:

1) Production of toxins — Many bacteria **release toxins** (harmful molecules) into the body. For example, the bacterium that causes **tetanus** produces a toxin that **blocks** the function of certain **nerve cells**, causing **muscle spasms.**
2) Cell damage — Pathogens can physically damage host cells by:
   - **Rupturing** them to **release nutrients** (proteins etc.) inside them.
   - **Breaking down** nutrients inside the cell for their own use. This starves and eventually **kills** the **cell.**
   - **Replicating** inside the cells and **bursting** them when they're released, e.g. some **viruses** do this.

The cell the pathogen has invaded and is reproducing inside is called the host cell.

## Lifestyle can Affect Your Risk of Developing Some Diseases

**1** Coronary heart disease (**CHD**) is a disease that affects your **heart** (see p. 44 for more). There are plenty of lifestyle factors that **increase** your **risk** of developing CHD:

1) **Poor diet** — a diet high in **saturated fat** or **salt** increases the risk.
2) **Smoking**, **lack of exercise** and **excessive alcohol intake** — these can all lead to **high blood pressure**, which can **damage** the heart and the blood vessels, **increasing** the risk of CHD

See p. 44 for more on risk factors and CHD.

**2** Cancer is the result of **uncontrolled cell division** (see p. 65). Factors that **increase** the **risk** of developing cancer include:

1) **Smoking** — the main cause of **mouth, throat** and **lung cancer** is smoking.
2) **Excessive exposure to sunlight** — excessive exposure can cause **skin cancer.** Using **sunbeds** and sunbathing **without sunscreen** increases the risk.
3) **Excessive alcohol intake** — this can increase the risk of many types of cancer, especially **liver cancer.**

# Disease

## It's never Too Late to Change Your Lifestyle

**Changing** your **lifestyle** for the **better** (e.g. reducing your alcohol intake, doing more exercise, eating healthily etc.) doesn't mean you'll **never** develop these diseases, but it can **reduce the risk**. So it's never too late to change. For example, studies have shown that the risk to a **smoker** of developing **lung cancer** is reduced if they stop smoking.

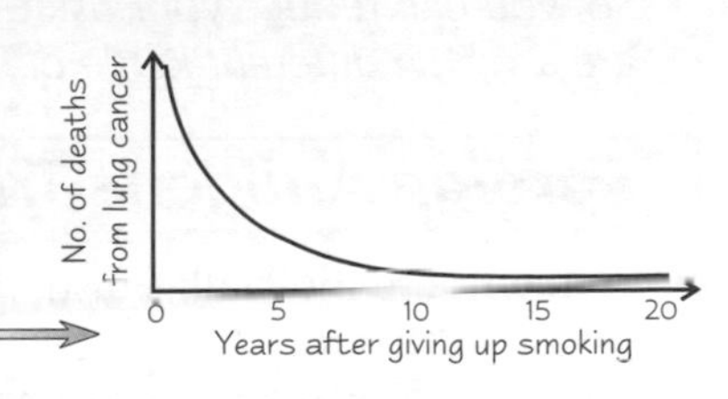


## You Need to be Able to Interpret Data About Risk Factors and Disease

In the exam you could be asked to **interpret data** about **lifestyle** and the **risk** of disease. Here's an example for you:

A study looked at the **link** between **body mass index** (BMI) and the **relative risk of developing cancer**. BMI is a measure of **obesity**. At the start of the study the BMI of **1.2 million women** aged 50-64 was taken, and after **five years** the **number** of these women **with cancer** was recorded. The relative risk of developing cancer was then worked out by taking into account **other factors** about the women, including **daily alcohol intake**, **smoking status** and **physical activity**. The graph on the right shows the results.

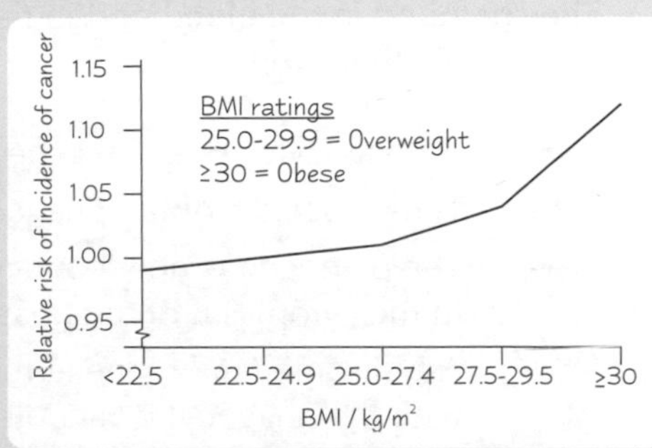


1) You might be asked to **describe the data...**
   The graph shows a **positive correlation** between **BMI** and the relative **risk** of cancer in women. The **higher** the BMI, the **higher** the risk of developing cancer. It's **not** a linear (straight-line) relationship though — the **risk increases** much more **quickly** for women with a **BMI over 27.5**.
2) ...or **draw conclusions**
   - The relative risk of developing cancer is **greatly increased** for women who are **overweight** or **obese**.
   - Be careful — you **can't conclude** that obesity **causes** cancer, only that they're **linked**. Another factor could be involved. For example, obese people are more likely to have **diets high in saturated fat** — it may be the saturated fat that causes the higher risk of cancer, not the actual increase in body mass.
   - You **can't conclude** that the risk of developing a **specific form** of cancer (e.g. throat cancer) increases with increasing BMI. The data doesn't deal with specific forms of cancer separately. So you can only conclude that the risk of developing cancer in **general** increases with increasing BMI.
3) ... or **evaluate the methodology**
   - The **sample size** is **large** — **1.2 million women**. This makes the results **more reliable**.
   - The study took into account **other lifestyle factors** (e.g. alcohol intake, smoking) that can **affect** the risk of developing cancer too. This also makes the results **more reliable**.

See pages 90-92 for more about interpreting data.

## Practice Questions

Q1 State four lifestyle factors that can affect your chances of getting CHD.

Q2 If you give up smoking, what happens to your risk of developing lung cancer?

**Exam Questions**

Q1 Describe three ways in which a pathogen may damage host cells. [3 marks]

Q2 Sketch a graph to show the likely correlation between the amount of time spent sunbathing and the incidence of skin cancer. [2 marks]

## Pizza, beer and telly a bad lifestyle? — depends who you ask...

*Phew, they were a tough two pages to start with, but they're done now and it's all fun, fun, fun from here on in. There's loads more interpreting data to come, so 'describe, draw conclusions and evaluate the methodology' might get a bit repetitive — but those pesky AQA examiners love asking you to interpret, so it's all good practice.*

# The Immune System

*So, you can reduce your risk of getting some non-infectious diseases by changing your lifestyle. But infectious diseases are a whole different kettle of fish. Luckily we have an army of cells that help to protect us — the immune system.*

## Foreign Antigens Trigger an Immune Response

**Antigens** are **molecules** (usually proteins or polysaccharides) found on the **surface** of **cells**. When a pathogen invades the body, the antigens on its cell surface are **identified as foreign**, which activates cells in the immune system. There are **four** main stages involved in the immune response:

## 1) Phagocytes Engulf Pathogens

A **phagocyte** (e.g. a macrophage) is a type of **white blood cell** that carries out **phagocytosis** (engulfment of pathogens). They're found in the **blood** and in **tissues** and are the first cells to respond to a pathogen inside the body. Here's how they work:

1) A phagocyte **recognises** the **antigens** on a pathogen.
2) The cytoplasm of the phagocyte moves round the pathogen, **engulfing** it.
3) The pathogen is now contained in a **phagocytic vacuole** (a bubble) in the cytoplasm of the phagocyte.
4) A **lysosome** (an organelle that contains **lysosomal enzymes**) **fuses** with the phagocytic vacuole. The lysosomal enzymes **break down** the pathogen.
5) The phagocyte then **presents** the pathogen's antigens — it sticks the antigens on its **surface** to **activate** other immune system cells.

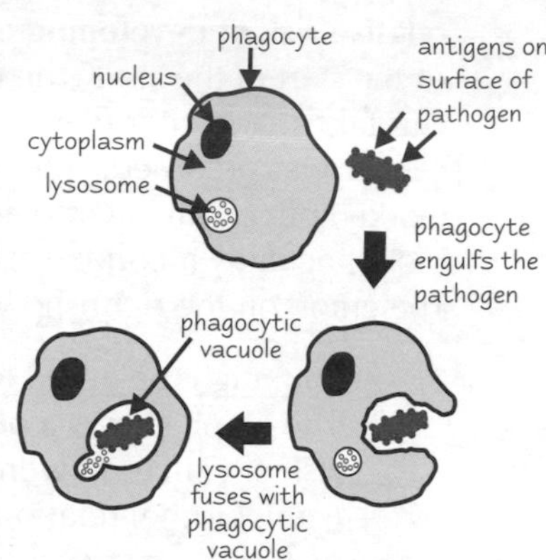


## 2) Phagocytes Activate T-cells

A **T-cell** is another type of **white blood cell**. It has **proteins** on its surface that **bind** to the **antigens** presented to it by **phagocytes**. This **activates** the T-cell. Different types of T-cells respond in different ways:

1) Some **release substances** to **activate B-cells**.
2) Some **attach** to antigens on a pathogen and **kill** the cell.

## 3) T-cells Activate B-cells, Which Divide into Plasma Cells

**B-cells** are also a type of **white blood cell**. They're covered with **antibodies** — proteins that **bind antigens** to form an **antigen-antibody complex**. Each B-cell has a **different shaped antibody** on its membrane, so different ones bind to **different shaped antigens**.

1) When the antibody on the surface of a B-cell meets a **complementary shaped** antigen, it binds to it.
2) This, together with substances released from T-cells, **activates** the B-cell.
3) The activated B-cell **divides** into **plasma cells**.

## 4) Plasma Cells Make More Antibodies to a Specific Antigen

**Plasma cells** are **identical** to the B-cell (they're **clones**). They secrete loads of the **antibody** specific to the antigen. Antibody **functions** include:

1) Coating the pathogen to make it easier for a **phagocyte** to engulf it.
2) Coating the pathogen to **prevent** it from **entering** host cells.
3) **Binding to** and **neutralising** (inactivating) toxins produced by the pathogen.

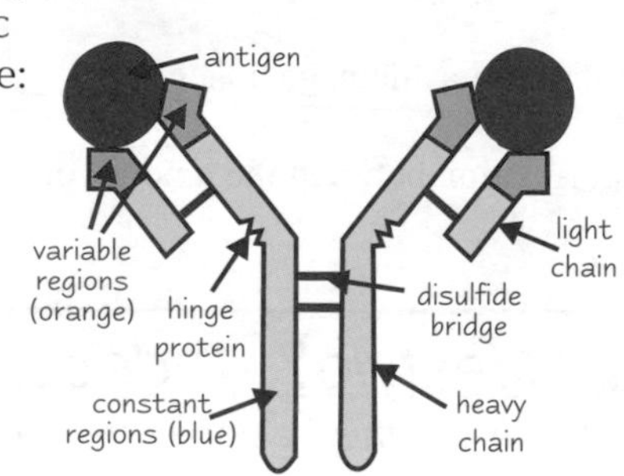


Antibodies are **proteins** — they're made up of chains of **amino acid** monomers linked by **peptide bonds** (see p. 14 for more on proteins). The **specificity** of an antibody depends on its **variable regions**. Each antibody has a **different shaped** variable region (due to different **amino acid sequences**) that's **complementary** to one **specific antigen**. The **constant regions** are the **same** in all antibodies.

# The Immune System

## The *Immune Response* Can be Split into *Cellular* and *Humoral*

Just to add to your fun, the **immune response** is often split into **two** — the **cellular response** and the **humoral response**.

1) **Cellular** — The **T-cells** and **other** immune system **cells** that they **interact** with, e.g. phagocytes, form the cellular response.
2) **Humoral** — **B-cells** and the production of **antibodies** form the **humoral response.**

**Both** types of response are **needed** to remove a pathogen from the body and the responses **interact** with each other, e.g. T-cells help to **activate** B-cells, and antibodies **coat** pathogens making it **easier** for **phagocytes** to **engulf** them.

## The *Immune Response* for Antigens can be *Memorised*

Neil's primary response — to his parents.

### The **Primary Response**

1) When an antigen enters the body for the **first time** it activates the immune system. This is called the **primary response**.
2) The primary response is **slow** because there **aren't many B-cells** that can make the antibody needed to bind to it.
3) Eventually the body will produce **enough** of the right antibody to overcome the infection. Meanwhile the infected person will show **symptoms** of the disease.
4) After being exposed to an antigen, both T- and B-cells produce **memory cells**. These memory cells **remain in the body** for a **long** time. Memory T-cells remember the **specific antigen** and will recognise it a second time round. Memory B-cells record the specific **antibodies** needed to bind the antigen.
5) The person is now **immune** — their immune system has the **ability** to respond **quickly** to a 2nd infection.

### The **Secondary Response**

1) If the **same pathogen** enters the body again, the immune system will produce a **quicker**, **stronger** immune response — the **secondary response.**
2) **Memory B-cells** divide into **plasma cells** that produce the right antibody to the antigen. **Memory T-cells** divide into the **correct type of T-cells** to kill the cell carrying the antigen.
3) The secondary response often gets rid of the pathogen **before** you begin to show any **symptoms.**

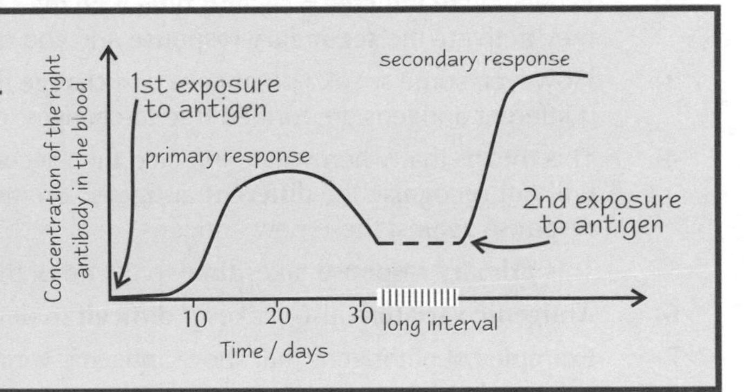


## Practice Questions

Q1 What are antigens?

Q2 What is phagocytosis?

Q3 What are the functions of T-cells and B-cells?

**Exam Questions**

Q1 Describe the function of antibodies. [3 marks]

Q2 Describe and explain how a secondary immune response differs to a primary immune response. [6 marks]

## *Memory cells — I need a lot more to cope with these pages...*

*If memory cells are mentioned in the exam remember that they are still types of T-cells and B-cells. They just hang around a lot longer than most T-cells and B-cells. When the antigen enters the body for a second time they can immediately divide into more of the specific T-cells and B-cells that can kill the pathogen or release antibodies against it.*

# Vaccines and Antibodies in Medicine

*The primary response gives us immunity against a disease, but only after you've become infected. If only there was a way to stimulate memory cell production without getting the disease... well, there is — vaccination.*

## *Vaccines* can *Protect Individuals* and *Populations* Against *Disease*

1) While your B-cells are busy **dividing** to build up their numbers to deal with a pathogen (i.e. the **primary response** — see previous page), you **suffer** from the disease. **Vaccination** can help avoid this.
2) Vaccines **contain antigens** that cause your body to **produce memory cells** against a particular pathogen, **without** the pathogen **causing disease**. This means you become **immune** without getting any **symptoms**.
3) Vaccines protect individuals that have them and, because they reduce the **occurrence** of the disease, those **not** vaccinated are also less likely to catch the disease (because there are fewer people to catch it from). This is called **herd immunity**.
4) Vaccines always contain antigens — these may be **free** or attached to a **dead** or **attenuated** (weakened) **pathogen**.
5) Vaccines may be **injected** or taken **orally**. The **disadvantages** of taking a vaccine orally are that it could be **broken down** by **enzymes** in the gut or the **molecules** of the vaccine may be **too large** to be **absorbed** into the blood.
6) Sometimes **booster** vaccines are given later on (e.g. after several years) to **make sure** that memory cells are produced.

The oral vaccine was proving hard to swallow.

## *Antigenic Variation* Helps Some *Pathogens Evade* the *Immune System*

1) **Antigens** on the surface of pathogens **activate** the **primary response**.
2) When you're **infected** a **second time** with the **same pathogen** (which has the **same antigens** on its surface) they **activate** the **secondary response** and you don't get ill.
3) However, some sneaky pathogens can **change** their surface antigens. This is called **antigenic variation**. (Different antigens are formed due to changes in the **genes** of a pathogen.)
4) This means that when you're infected for a **second time**, the **memory cells** produced from the **first infection** will **not recognise** the **different antigens**. So the immune system has to start from scratch and carry out a **primary response** against these new antigens.
5) This **primary response** takes **time** to get rid of the infection, which is why you get **ill again**.
6) **Antigenic variation** also makes it **difficult** to develop **vaccines** against some pathogens for the same reason.
7) **Examples** of pathogens that show antigenic variation include **HIV**, ***S. pneumoniae*** bacteria and the **influenza virus**. You need to **learn** how it works in influenza:

**Antigenic variation in the influenza virus**

1) The **influenza virus** causes **influenza** (flu).
2) **Proteins** (**neuraminidase** and **haemagglutinin**) on the **surface** of the influenza virus act as **antigens**, **triggering** the immune system.
3) These antigens can **change regularly**, forming **new strains** of the virus.
4) **Memory cells** produced from **infection** with **one strain** of flu will **not recognise** other strains with **different antigens**.
5) This means your immune system produces a **primary response** every time you're infected with a **new strain** (carrying different antigens).
6) So this means you can **suffer from flu** more than once — each time you're infected with a **new strain**.

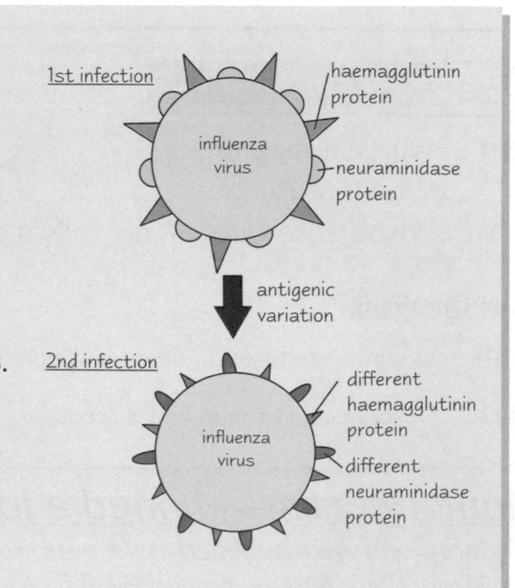

# Vaccines and Antibodies in Medicine

## *Monoclonal Antibodies* can be used to *Target Specific Substances* or *Cells*

1) **Monoclonal antibodies** are antibodies **produced** from a **single group of genetically identical B-cells** (plasma cells). This means that they're all **identical** in **structure**.
2) As you know, antibodies are **very specific** because their binding sites have a **unique structure** that only one particular antigen will fit into (one with a **complementary shape**).
3) You can make monoclonal antibodies **that bind to anything** you want, e.g. a cell antigen or other substance, and they will only bind to (target) this molecule.

**EXAMPLE: TARGETING CELLS — CANCER**

1) **Different cells** in the body have **different** surface **antigens**.
2) Cancer cells have antigens called **tumour markers** that are **not** found on normal body cells.
3) **Monoclonal antibodies** can be made that will bind to the tumour markers.
4) You can also attach **anti-cancer drugs** to the antibodies.
5) When the antibodies come into **contact** with the cancer cells they will **bind** to the tumour markers.
6) This means the drug will **only accumulate** in the body where there are **cancer cells**.
7) So, the **side effects** of an antibody-based drug are lower than other drugs because they accumulate near **specific cells**.

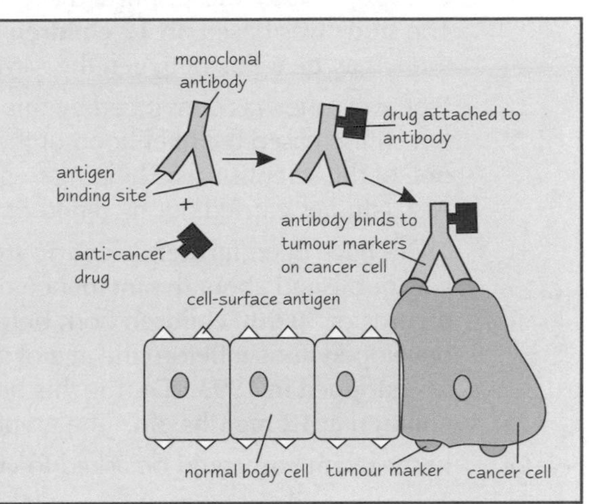


**EXAMPLE: TARGETING SUBSTANCES — PREGNANCY TESTS**

Pregnancy tests detect the hormone **human chorionic gonadotropin (hCG)** that's found in the **urine** of pregnant women:

1) The application area contains **antibodies for hCG** bound to a **coloured bead** (**blue**).
2) When urine is applied to the application area any hCG will **bind** to the antibody on the beads, forming an **antigen-antibody complex**.
3) The urine **moves** up the stick to the **test strip**, **carrying** any **beads** with it.
4) The test strip contains **antibodies to hCG** that are stuck in place (**immobilised**).
5) If there **is hCG present** the test strip turns **blue** because the **immobilised** antibody binds to any **hCG** — concentrating the hCG-antibody complex with the **blue beads** attached. If **no hCG** is present, the beads will **pass through** the test area **without** binding to anything, and so it **won't** go blue.

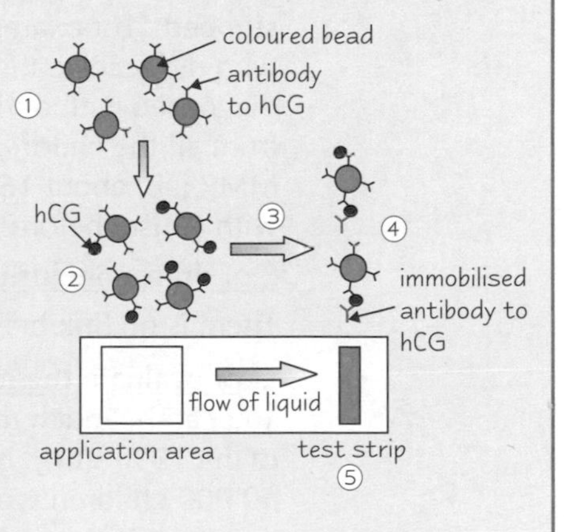


## *Practice Questions*

Q1 How do vaccines cause immunity?

Q2 What are monoclonal antibodies?

**Exam Questions**

Q1 Explain why it is possible to suffer from the flu more than once. [4 marks]

Q2 Describe how monoclonal antibodies can be used to target a drug to cancer cells. [4 marks]

## *An injection of dead bugs — roll on my next vaccine...*

*Monoclonal antibodies are really useful — they can even be made against other antibodies. For example, people with asthma produce too many of a type of antibody that causes inflammation in the lungs. Monoclonal antibodies can be made to bind this type of antibody, so it can no longer cause inflammation, which can reduce the symptoms of asthma sufferers.*

# Interpreting Vaccine and Antibody Data

*If someone claims anything about a vaccine or an antibody the claim has to be validated (confirmed) before it's accepted.*

## ***New Knowledge** About **Vaccines** and **Antibodies** is **Validated** by **Scientists***

When a **study** presents evidence for a **new theory** (e.g. that a vaccine has a dangerous side effect) it's important that other scientists come up with **more evidence** in order to **validate** (confirm) the theory. To validate the theory other scientists may **repeat** the study and try to **reproduce** the results, or **conduct other studies** to try to prove the same theory (see p. 2).

**EXAMPLE 1: The MMR Vaccine**

1) In 1998, a study was published about the **safety** of the **measles**, **mumps** and **rubella (MMR) vaccine**. The study was based on **12 children** with **autism** (a life-long developmental disability) and concluded that there may be a **link** between the MMR vaccine and autism.
2) Not everyone was convinced by this study because it had a **very small sample size** of 12 children, which increased the likelihood of the results being due to **chance**. The study may have been **biased** because one of the scientists was helping to gain evidence for a **lawsuit** against the MMR vaccine manufacturer. Also, studies carried out by different scientists found no link between autism and the MMR vaccine.
3) There have been **further scientific studies** to sort out the **conflicting** evidence. In **2005**, a **Japanese** study was published about the incidence of autism in Yokohama (an area of Japan). They looked at the medical records of **30 000 children** born between **1988 and 1996** and counted the number of children that developed **autism** before the age of seven. The **MMR jab** was first **introduced in Japan in 1989** and was **stopped in 1993**. During this time the MMR vaccine was administered to children at **12 months old**. The graph shows the results of the study.
4) In the exam you could be asked to **evaluate evidence** like this.

See pages 90-92 for more about evaluating data.

- <u>You might be asked to explain the data...</u>
  The graph shows that the number of children diagnosed with autism continued to **rise** after the MMR vaccine was **stopped**. For example, from all the children born in 1992, who did receive the MMR jab, about 60 out of 10 000 were diagnosed with autism before the age of seven. However, from all the children born in 1994, who did not receive the MMR jab, about 160 out of 10 000 of them were diagnosed with autism before the age of seven.
- <u>...or draw conclusions</u>
  There is **no link** between the MMR vaccine and autism.
- <u>... or evaluate the methodology</u>
  You can be much more confident in this study, compared to the 1998 study, because the **sample size** was so **large** — 30 000 children were studied. A larger sample size means that the results are less likely to be due to **chance**.

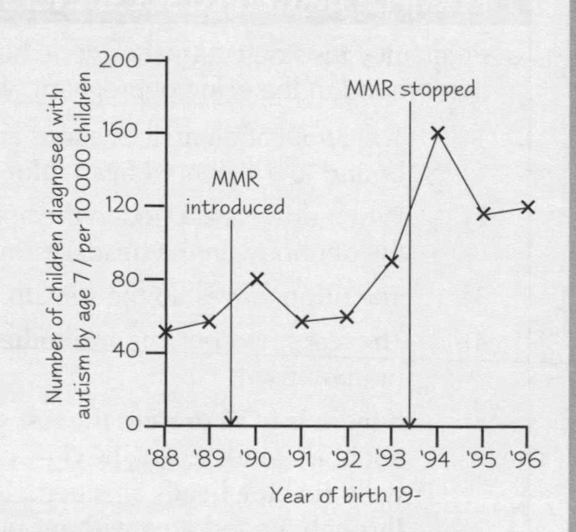


**EXAMPLE 2: Herceptin — Monoclonal Antibodies**

About **20%** of **women with breast cancer** have tumours that produce more than the usual amount of a **receptor** called **HER2**. **Herceptin** is a **drug** used to treat this type of breast cancer — it contains **monoclonal antibodies** that **bind the HER2 receptor** on a **tumour cell** and **prevent** the cells from growing and dividing.

In **2005**, a study **tested** Herceptin on women who had already undergone **chemotherapy** for HER2-type **breast cancer**. **1694** women took the **drug** for a **year** after chemotherapy and another **1694** women were **observed** for the **same time** (the control group). The results are shown in the graph on the right.

**<u>Describe the data:</u>** Almost **twice as many** women in the **control group** developed breast cancer again or died **compared** to the group taking Herceptin.

**<u>Draw conclusions:</u>** A **one-year treatment** with Herceptin, after chemotherapy, **increases** the disease-free survival rate for women with HER2-type breast cancer.

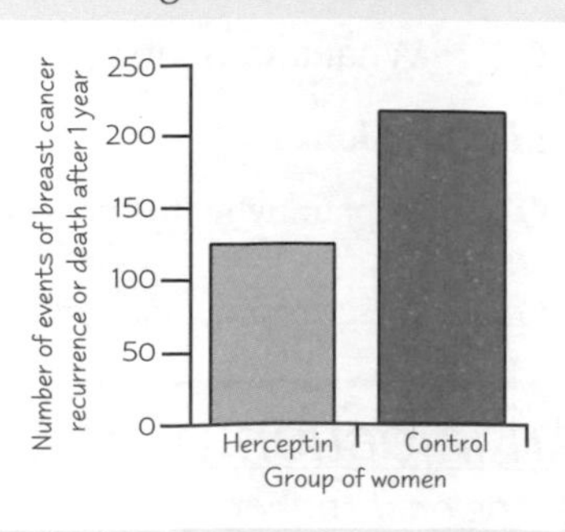

# Interpreting Vaccine and Antibody Data

## We use Scientific Knowledge to Make Decisions

When **new scientific information** about **vaccines** and **monoclonal antibodies** has been **validated** by scientists, **society** (organisations and the public) can **use** this information to make **informed decisions**. **Two examples** are given below:

**EXAMPLE 1: The MMR Vaccine**

Scientific knowledge:

The **validity** of the 1998 study that linked MMR and autism is in doubt. **New studies** have shown **no link** between the vaccine and autism.

Decision:

Scientists and doctors still recommended that parents **immunise** their **children** with the **MMR vaccine**.

**EXAMPLE 2: Herceptin — Monoclonal Antibodies**

Scientific knowledge:

**Early studies** about Herceptin showed **severe heart problems** could be a **side effect** of the drug.

Decision:

All patients receiving Herceptin must be **monitored** for heart problems, e.g. by having **heart tests** done.

## Use of Vaccines and Antibodies Raises Ethical Issues

**Ethical issues surrounding vaccines include:**

1) All vaccines are **tested on animals** before being tested on humans — some people **disagree** with animal testing. Also, **animal based substances** may be used to **produce** a vaccine, which some people disagree with.
2) **Testing** vaccines on **humans** can be **tricky**, e.g. volunteers may put themselves at **unnecessary risk** of contracting the disease because they think they're fully protected (e.g. they might have unprotected sex because they have had a new HIV vaccine and think they're protected — and the vaccine might not work).
3) Some people **don't** want to take the vaccine due to the **risk** of **side effects**, but they are **still protected** because of **herd immunity** (see p. 8) — other people think this is **unfair**.
4) If there was an **epidemic** of a **new disease** (e.g. a new influenza virus) there would be a rush to **receive** a vaccine and **difficult decisions** would have to be made about **who** would be the **first** to receive it.

**Ethical issues surrounding monoclonal antibody therapy** often involve animal rights issues. **Animals** are used to **produce the cells** from which the monoclonal antibodies are produced. Some people **disagree** with the use of animals in this way.

## Practice Questions

Q1 Suggest one ethical issue surrounding vaccines.

Q2 Suggest one ethical issue surrounding monoclonal antibodies.

**Exam Question**

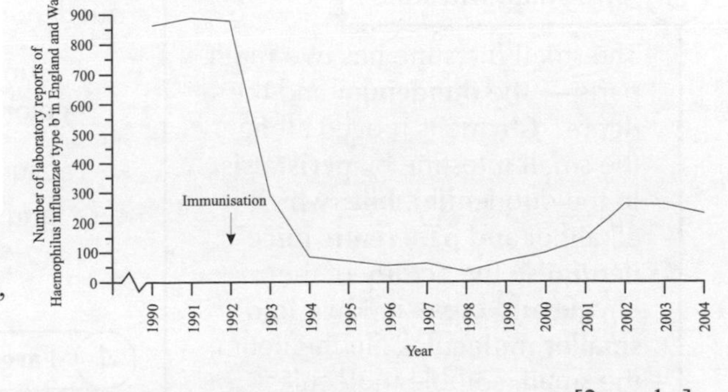


Q1 The graph on the right shows the number of laboratory reports of *Haemophilus influenzae* type b (Hib), in England and Wales, from 1990 to 2004. Hib affects children and can lead to meningitis and pneumonia.

a) Why did the number of cases of Hib decrease after 1992? [2 marks]

b) Due to a shortage of the normal vaccine in 2000-2001, a different type of Hib vaccine was used. What effect did this have on the number of cases of Hib? [1 mark]

## Some scientists must have to validate the taste of chocolate — nice job...

*After the 1998 study, some parents were worried about giving their kids the MMR vaccine, so the number of children given the vaccine fell. With fewer children in each community protected by the vaccine, herd immunity decreased. This meant that more people were vulnerable to the diseases, so the number of cases of measles, mumps and rubella went up.*

# The Digestive System

*With the digestive system in contact with the environment and us left open to pathogens, eating can be a pretty risky game. It's dead important though — without it we'd lack all the energy and important nutrients we need.*

## Digestion Breaks Down Large Molecules into Smaller Molecules

1) Many of the molecules in our **food** are **polymers**.
2) These are **large**, complex molecules composed of long chains of **monomers** — small **basic molecular units**.
3) **Proteins** and some **carbohydrates** are **polymers**. In carbohydrates, the monomers are called **monosaccharides**. They contain the elements **carbon**, **hydrogen** and **oxygen**. In proteins the monomers are called **amino acids**. They contain **carbon**, **hydrogen**, **oxygen**, **nitrogen**.

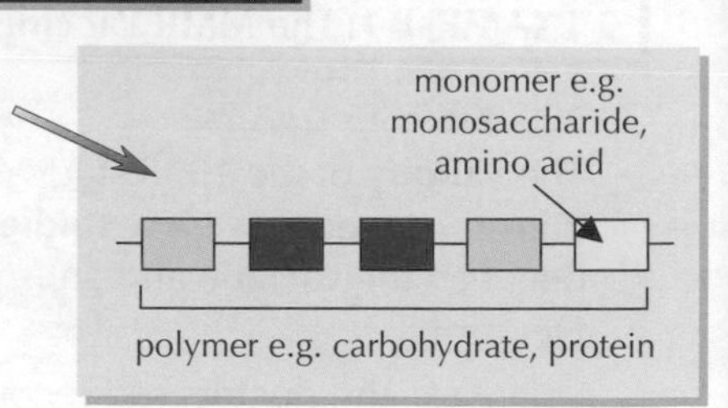


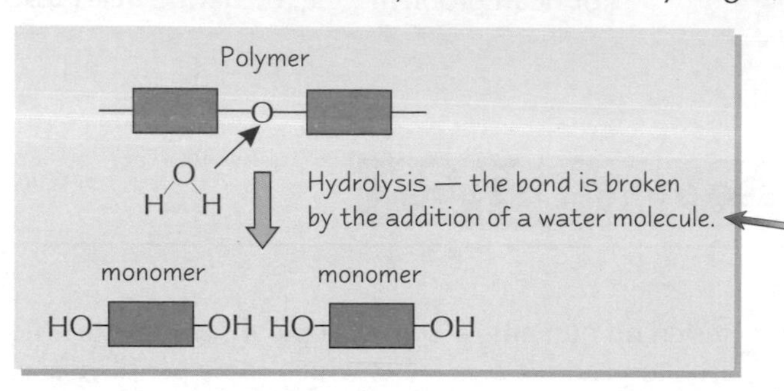


4) The polymers in our food are **insoluble** — they can't be directly **absorbed** into our bloodstream and **assimilated** (made) into new products.
5) The polymers have to be **hydrolysed** (broken down) into **smaller**, more **soluble** molecules by **adding water**.
6) This process happens during **digestion**.
7) **Hydrolysis** is catalysed by **digestive enzymes**.

## Each Part of the Digestive System Has a Specific Function

All the organs in the **digestive system** have a **role** in **breaking down** food and **absorbing nutrients**:

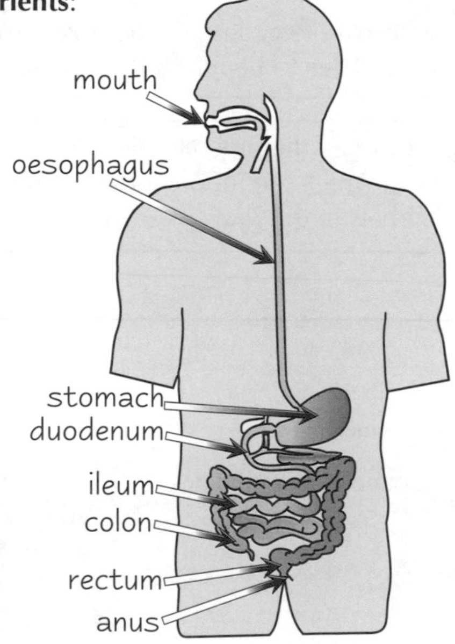


### 1 Oesophagus

The **tube** that takes **food** from the **mouth** to the **stomach** using waves of **muscle contractions** called **peristalsis**. **Mucus** is secreted from tissues in the walls, to **lubricate** the food's passage downwards.

### 2 The Stomach

The stomach is a **small sac**. It has lots of **folds**, allowing the stomach to **expand** — it can hold up to 4 litres of food and liquid. The entrance and exit of the stomach are controlled by **sphincter muscles**. The stomach walls produce **gastric juice**, which helps break down food. Gastric juice consists of **hydrochloric acid** (HCl), **pepsin** (an enzyme) and **mucus**. **Pepsin** hydrolyses **proteins**, into smaller polypeptide chains. It only works in **acidic conditions** (provided by the HCl). **Peristalsis** of the stomach turns food into an acidic fluid called **chyme**.

### 3 The Small Intestine

The small intestine has two main parts — the **duodenum** and the **ileum**. **Chyme** is moved along the small intestine by **peristalsis**. In the duodenum, **bile** (which is alkaline) and **pancreatic juice** **neutralise** the **acidity** of the chyme and break it down into **smaller molecules**. In the ileum, the small, soluble molecules (e.g. glucose and amino acids) are **absorbed** through structures called **villi** that line the gut wall. Molecules are absorbed by **diffusion**, **facilitated diffusion** and **active transport** (see pages 28-30 for more).

### 4 Large intestine

The large intestine (colon) absorbs **water**, **salts** and **minerals.** Like other parts of the digestive system, it has a **folded wall** — this provides a **large surface area** for absorption. **Bacteria** that **decompose** some of the undigested nutrients are found in the large intestine.

### 5 Rectum

**Faeces** are stored in the rectum and then pass through **sphincter** muscles at the **anus** during **defecation**. Nice.

# The Digestive System

## The Pancreas and Salivary Glands Play Important Roles in Digestion

**Glands** along the **digestive system** release **enzymes** to help **break down** food.
You need to know about two of these glands — the **salivary glands** and the **pancreas**.

### The Salivary Glands

There are three main pairs of salivary glands in the mouth. They secrete **saliva** that consists of **mucus**, **mineral salts** and **salivary amylase** (an enzyme). Salivary amylase breaks down **starch** into **maltose**, a disaccharide (see p. 16). Saliva has other roles in digestion — e.g. it helps to **lubricate** food, making it easier to **swallow**.

### The Pancreas

The pancreas releases **pancreatic juice** into the **duodenum** (see previous page) through the **pancreatic duct**. Pancreatic juice contains **amylase**, **trypsin**, **chymotrypsin** and **lipase** (the functions of these enzymes are listed below). It also contains **sodium hydrogencarbonate**, which **neutralises** the acidity of **hydrochloric acid** from the **stomach**.

## Enzymes Help us to Digest Food Molecules

Digestive enzymes can be divided into three classes.

1) **Carbohydrases** catalyse the hydrolysis of **carbohydrates**.
2) **Proteases** catalyse the hydrolysis of **proteins**.
3) **Lipases** catalyse the hydrolysis of **lipids**.

The table shows some more specific enzyme reactions.

| LOCATION | ENZYME | CLASS | HYDROLYSES | INTO |
|---|---|---|---|---|
| salivary glands | **amylase** | carbohydrase | starch | maltose |
| stomach | **pepsin** | protease | protein | peptides |
| pancreas | **amylase** | carbohydrase | starch | maltose |
| | **trypsin** | protease | protein | peptides |
| | **chymotrypsin** | protease | protein | peptides |
| | **carboxypeptidase** | protease | peptides | amino acids |
| | **lipase** | lipase | lipids | fatty acids + glycerol |
| ileum | **maltase** | carbohydrase | maltose | glucose |
| | **sucrase** | carbohydrase | sucrose | glucose + fructose |
| | **lactase** | carbohydrase | lactose | glucose + galactose |
| | **peptidase** | protease | peptides | amino acids |

## Practice Questions

Q1 What is hydrolysis?

Q2 Describe the structure and function of the stomach.

Q3 Describe the structure and function of the small intestine.

Q4 What do lipases break down?

**Exam Question**

Q1 There are several glands associated with the digestive system.

a) Describe how the pancreas aids digestion. [6 marks]

b) Name one other gland associated with the digestive system. [1 mark]

## Gastric juice — not for me thanks, think I'll stick to orange...

*So when I eat a biscuit, first it gets chomped up by my teeth. Then it's broken down by carbohydrases and lipases, before all those smaller, more soluble molecules are absorbed and, most likely, deposited on my hips. Then there's the messy business of defecation. Learn all this stuff on the digestive system inside out — it might just save your life.*

# Proteins

*Protein is found in loads of different foods like chicken, eggs and nuts. But what are proteins? How are they made? What do they look like? Well, for your enjoyment, here are the answers to all those questions and many, many more...*

## ***Proteins*** *are Made from* ***Long Chains*** *of* ***Amino Acids***

1) The **monomers** of proteins are **amino acids**.
2) A **dipeptide** is formed when **two** amino acids join together.
3) A **polypeptide** is formed when **more than two** amino acids join together.
4) **Proteins** are made up of **one or more polypeptides**.

Grant's cries of "die peptide, die" could be heard for miles around. He'd never forgiven it for sleeping with his wife.

## ***Different Amino Acids*** *Have* ***Different Variable Groups***

All amino acids have the same general structure — a **carboxyl group** (-COOH) and an **amino group** ($-NH_2$) attached to a **carbon** atom. The **difference** between different amino acids is the **variable** group (**R** on diagram) they contain.

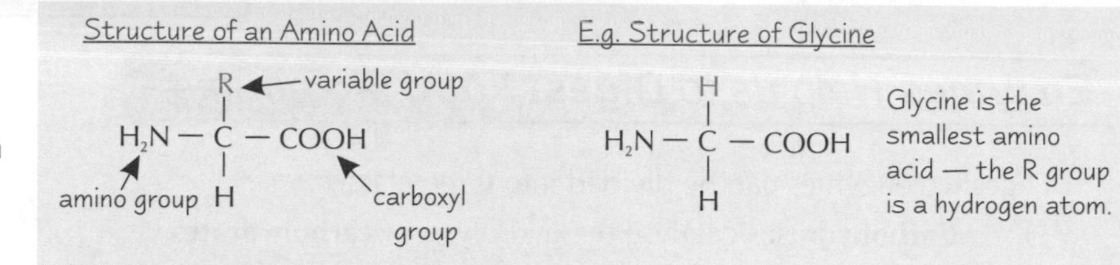


## ***Polypeptides*** *are Formed by* ***Condensation Reactions***

Amino acids are linked together by **condensation** reactions to form polypeptides. A molecule of **water** is **released** during the reaction. The bonds formed between amino acids are called **peptide bonds**. The reverse reaction happens during digestion.

amino acid 1
amino acid 2
condensation
hydrolysis
dipeptide
$H_2O$
a molecule of water is formed during condensation.
peptide bond

## ***Proteins*** *Have* ***Four Structural Levels***

Proteins are **big**, **complicated** molecules. They're much easier to explain if you describe their structure in four 'levels'. These levels are a protein's **primary**, **secondary**, **tertiary** and **quaternary** structures.

**Primary Structure** — this is the **sequence** of **amino acids** in the **polypeptide chain**.

**Secondary Structure** — the polypeptide chain doesn't remain flat and straight. **Hydrogen bonds** form between the amino acids in the chain. This makes it automatically **coil** into an **alpha (α) helix** or **fold** into a **beta (β) pleated sheet** — this is the secondary structure.

**Tertiary Structure** — the coiled or folded chain of amino acids is often **coiled** and **folded further**. **More bonds** form between different parts of the polypeptide chain. For proteins made from a **single** polypeptide chain, the tertiary structure forms their **final 3D structure**.

**Quaternary Structure** — some proteins are made of **several different polypeptide chains** held together by **bonds**. The **quaternary structure** is the way these polypeptide chains are assembled together. For proteins made from more than one polypeptide chain (e.g. haemoglobin, insulin, collagen), the quaternary structure is the protein's **final 3D structure**.

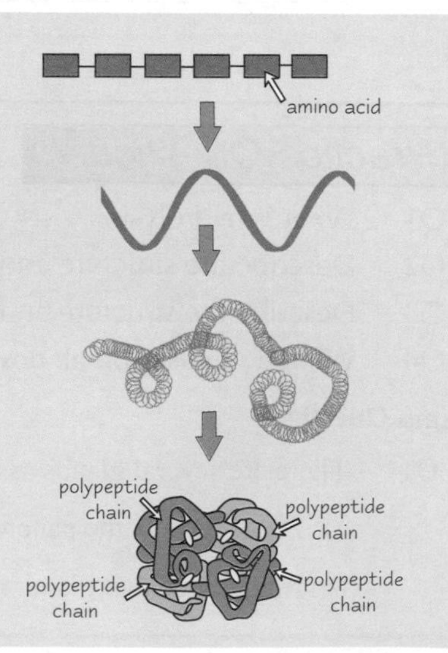


A protein's **shape** determines its **function**. E.g. **haemoglobin** is a **compact**, **soluble protein**, which makes it easy to **transport**. This makes it great for **carrying oxygen** around the body (see p. 58). **Collagen** has three polypeptide chains **tightly coiled** together, which makes it **strong**. This makes it a great **supportive tissue** in animals.

# Proteins

## Proteins have a Variety of Functions

There are **loads** of different **proteins** found in **living organisms**. They've all got **different structures** and **shapes**, which makes them **specialised** to carry out particular **jobs**. For example:

1) **Enzymes** — they're usually roughly **spherical** in shape due to the **tight folding** of the polypeptide chains. They're **soluble** and often have roles in **metabolism**, e.g. some enzymes break down large food molecules (**digestive enzymes**, see p. 13) and other enzymes help to **synthesise** (make) large molecules.

2) **Antibodies** — are involved in the **immune response**. They're made up of **two light** (short) polypeptide chains and **two heavy** (long) polypeptide chains bonded together. Antibodies have **variable regions** (see p. 6) — the **amino acid sequences** in these regions **vary** greatly.

3) **Transport proteins** — are present in **cell membranes** (p. 29). They contain **hydrophobic** (water hating) and **hydrophilic** (water loving) amino acids, which cause the protein to **fold up** and form a **channel**. These proteins **transport molecules** and **ions** **across** membranes.

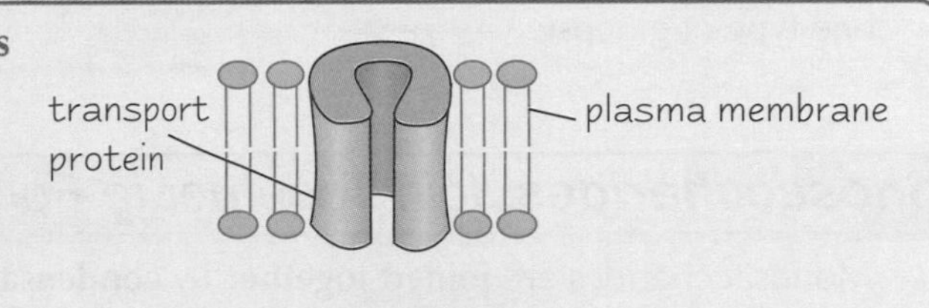


4) **Structural proteins** — are physically **strong**. They consist of **long polypeptide chains** lying **parallel** to each other with **cross-links** between them. Structural proteins include **keratin** (found in hair and nails) and **collagen** (found in connective tissue).

## Use the Biuret Test for Proteins

If you needed to find out if a substance, e.g. a **food sample**, contained **protein** you'd use the **biuret test**.

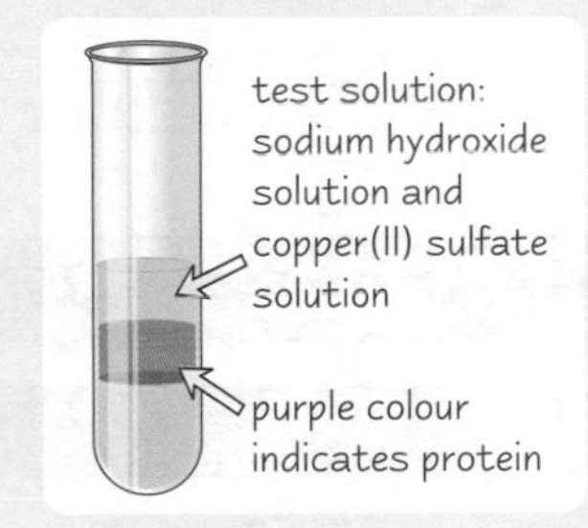


There are **two stages** to this test.

1) The test solution needs to be **alkaline**, so first you add a few drops of **sodium hydroxide solution**.
2) Then you add some **copper(II) sulfate solution**.

- If protein **is** present a **purple layer** forms.
- If there's **no protein**, the solution will **stay blue**. The colours are pale, so you need to look carefully.

## Practice Questions

Q1 What groups do all amino acid molecules have in common?

Q2 Give three functions of proteins.

Q3 Describe how you would test for the presence of protein in a sample.

**Exam Questions**

Q1 Describe how a dipeptide is formed. [5 marks]

Q2 Describe the structure of a protein, explaining the terms primary, secondary, tertiary and quaternary structure. No details of the chemical nature of the bonds are required. [9 marks]

## Condensation — I can see the reaction happening on my car windows...

*Protein structure is hard to imagine. I think of a Slinky — the wire's the primary structure, it coils up to form the secondary structure and if you coil the Slinky round your arm that's the tertiary structure. When a few Slinkies get tangled up that's like the quaternary structure. Oh, I need to get out more. I wish I had more friends and not just this stupid Slinky for company.*

# Carbohydrates

*Carbohydrates are present in foods like pasta, potatoes, bread and cakes — basically all of the yummy stuff. You need to know how they're made in the first place. So go grab a potato... and read on...*

## Carbohydrates are Made from **Monosaccharides**

1) As you know, most carbohydrates are **polymers**.
2) All carbohydrates contain the elements **C**, **H** and **O**.
3) The **monomers** that they're made from are **monosaccharides**, e.g. **glucose**, **fructose** and **galactose**. You need to learn the structure of one type of glucose:

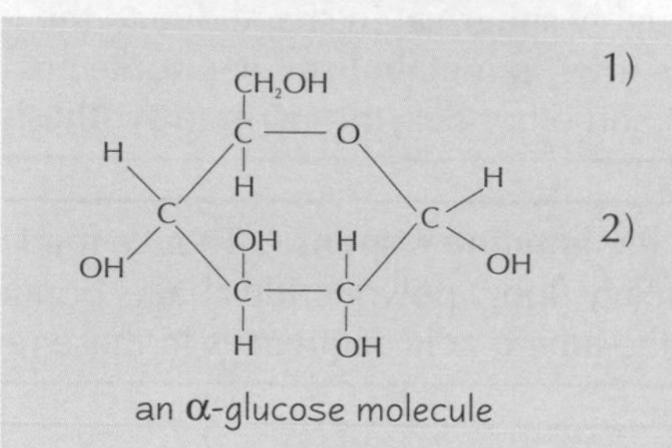

an α-glucose molecule

1) Glucose is a **hexose sugar** — a monosaccharide with **six carbon** atoms in each molecule.
2) There are two forms of glucose — **alpha** ($\alpha$) and **beta** ($\beta$) glucose (see p. 60 for more on β-glucose). For this Unit you need to learn the **structure** of α-glucose.

## **Monosaccharides** Join Together to Form **Disaccharides** and **Polysaccharides**

1) Monosaccharides are **joined together** by **condensation reactions**.
2) During a condensation reaction a molecule of **water** is **released** and a **glycosidic bond** forms between the two monosaccharides.
3) A **disaccharide** is formed when **two monosaccharides** join together. A **polysaccharide** is formed when **more than two monosaccharides** join together.

**Example**

Two **α-glucose** molecules are joined together by a **glycosidic bond** to form **maltose**.

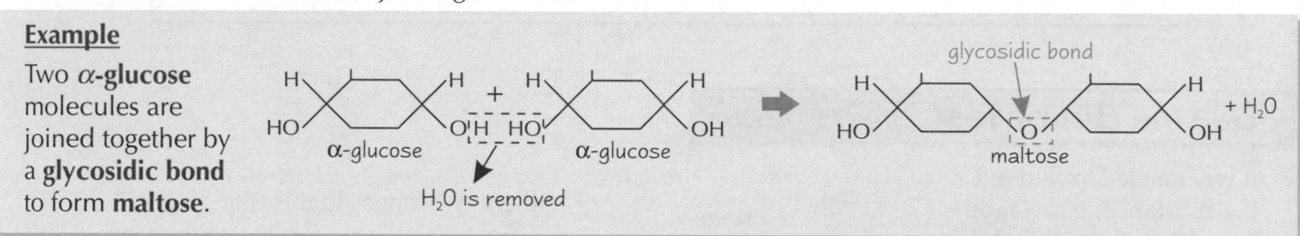


## **Disaccharides** and **Polysaccharides** are **Broken Down** During **Digestion**

Disaccharides and polysaccharides are often present in the **food** we eat, so we need to be able to break them down. Luckily we have **enzymes** released by the **intestinal epithelium** (see p. 13), that **hydrolyse** (break down) disaccharides and polysaccharides.

You need to learn the **monosaccharides** that make up **maltose**, **sucrose** and **lactose** and the **enzymes** that **hydrolyse** them. Luckily for you we've put it all in this pretty purple table.

| Disaccharide | Hydrolysed by... | Into... |
|---|---|---|
| maltose | maltase | glucose + glucose |
| sucrose | sucrase | glucose + fructose |
| lactose | lactase | glucose + galactose |

## **Lactose-Intolerance** is Caused by a **Lack** of the Digestive Enzyme **Lactase**

1) **Lactose** is a **sugar** found in milk.
2) It's digested by an **enzyme** called **lactase**, found in the intestines.
3) If you **don't** have enough of the enzyme lactase, you won't be able to break down the lactose in milk properly — a condition called **lactose-intolerance**.
4) Undigested lactose is fermented by bacteria and can cause a whole host of **intestinal complaints** such as **stomach cramps**, excessive **flatulence** (wind) and **diarrhoea**.
5) Milk can be artificially treated with purified lactase to make it suitable for lactose-intolerant people.
6) It's fairly **uncommon** to be lactose **tolerant** though — around 15% of Northern Europeans, 50% of Mediterraneans, 95% of Asians and 90% of people of African descent are lactose intolerant.

# Carbohydrates

## Use the **Benedict's Test** for **Sugars**

**Sugar** is a general term for **monosaccharides** and **disaccharides**. All sugars can be classified as **reducing sugars** or **non-reducing sugars**. If you carry out an **experiment** on the **digestion** of **carbohydrates** you'll need to **test** for sugars — to do this you use the **Benedict's test**. The test **differs** depending on the **type** of sugar you're testing for.

**REDUCING SUGARS**

1) Reducing sugars include **all monosaccharides** and **some disaccharides**, e.g. maltose.
2) You add **Benedict's reagent** (which is **blue**) to a sample and **heat it**. If the sample contains reducing sugars it gradually turns **brick red** due to the formation of a **red precipitate**.

**NON-REDUCING SUGARS**

1) To test for **non-reducing sugars**, like sucrose, first you have to break them down into monosaccharides.
2) You do this by **boiling** the test solution with **dilute hydrochloric acid** and then **neutralising** it with **sodium hydrogencarbonate**. Then just carry out the **Benedict's test** as you would for a reducing sugar.
3) Annoyingly, if the result of this test is **positive** the sugar could be reducing **or** non-reducing. To **check** it's non-reducing you need to do the **reducing sugar test** too (to rule out it being a reducing sugar).

## **Starch** is Made from **Two Polysaccharides**

1) Starch is made up of a mixture of two polysaccharides — **amylose** and **amylopectin** (see p. 60).
2) Both are composed of **long chains** of **α-glucose** linked together by glycosidic bonds, formed in condensation reactions.
3) When starch is digested, it's first broken down into **maltose** by **amylase** — an enzyme released by the **salivary glands** and the **pancreas** (see p. 13).
4) Maltose is then broken down into **α-glucose molecules** by **maltase**, which is released by the **intestinal epithelium**.

Jamelia would never even dream of leaving the house without her collar starched.

## Use the **Iodine Test** for **Starch**

If you do any **experiment** on the **digestion** of **starch** and want to find out if any is **left**, you'll need the **iodine test**.

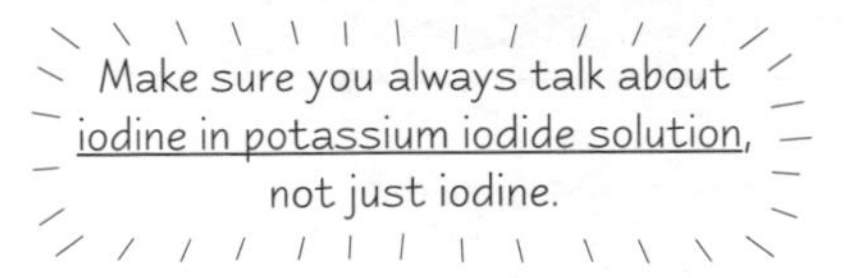


Just add **iodine dissolved in potassium iodide solution** to the test sample. If there **is starch present**, the sample changes from **browny-orange** to a dark, **blue-black** colour.

## Practice Questions

Q1 Draw the structure of α-glucose.

Q2 What type of bond holds monosaccharide molecules together in a disaccharide?

Q3 Name the enzymes that break down: i) sucrose, ii) maltose, iii) lactose.

Q4 Name the two polysaccharides present in starch.

**Exam Questions**

Q1 Describe the cause and symptoms of lactose intolerance. [4 marks]

Q2 Describe the test used to identify a non-reducing sugar.
Include the result you would expect to see if the test was positive. [6 marks]

## Reducing sugars — who on earth would want to do that?

*Just to confuse matters, in addition to α-glucose, there's also β-glucose. But you don't have to worry about that for now. It ain't fun learning the structure of glucose... you basically have to copy it down, then cover the page and test yourself until you know it off by heart. But at least then there'll be no nasty surprises in the exam...*

# Enzyme Action

*In our digestive system, enzymes help to break down all the stuff we eat. Without them we wouldn't get the nutrients and energy from our food very quickly at all... which wouldn't be good. So here's how they work....*

## Enzymes are Biological Catalysts

Enzymes **speed up chemical reactions** by acting as **biological catalysts**.

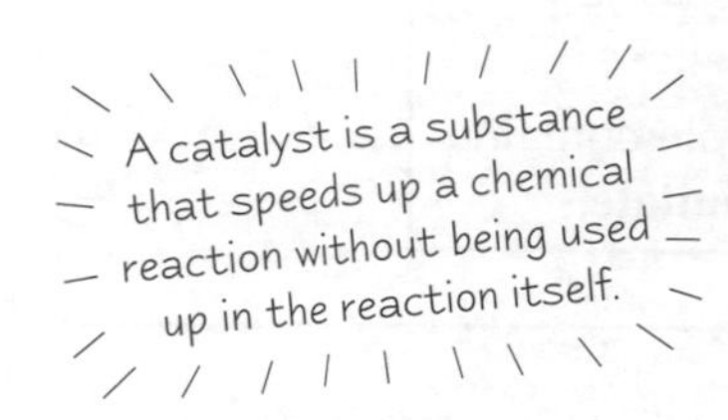


1) They catalyse **metabolic reactions** in your body, e.g. digestion and respiration. Even your **phenotype** (physical appearance) is due to enzymes that catalyse the reactions that cause growth and development (see p. 52).
2) Enzymes are **proteins** (see p. 15).
3) Enzymes have an **active site**, which has a **specific shape**. The active site is the part of the enzyme where the **substrate** molecules (the substance that the enzyme interacts with) **bind to**.
4) Enzymes are **highly specific** due to their tertiary structure (see next page).

## Enzymes Lower the Activation Energy of a Reaction

In a chemical reaction, a certain amount of **energy** needs to be supplied to the chemicals before the reaction will **start**. This is called the **activation energy** — it's often provided as **heat**. Enzymes **lower** the amount of activation energy that's needed, often making reactions happen at a **lower temperature** than they could without an enzyme. This **speeds up** the **rate of reaction**.

When a substrate fits into the enzyme's active site it forms an **enzyme-substrate complex** — it's this that lowers the activation energy. Here are two reasons why:

1) If two substrate molecules need to be **joined**, being attached to the enzyme holds them **close together**, **reducing** any **repulsion** between the molecules so they can bond more easily.
2) If the enzyme is catalysing a **breakdown reaction**, fitting into the active site puts a **strain** on bonds in the substrate, so the substrate molecule **breaks up** more easily.

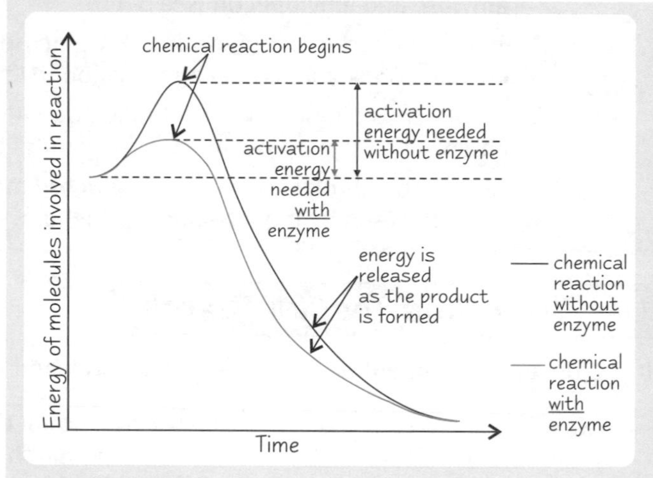


## The 'Lock and Key' Model is a Good Start...

Enzymes are a bit picky — they only work with substrates that fit their active site. Early scientists studying the action of enzymes came up with the '**lock and key**' model. This is where the **substrate fits** into the **enzyme** in the same way that a **key fits** into a **lock**.

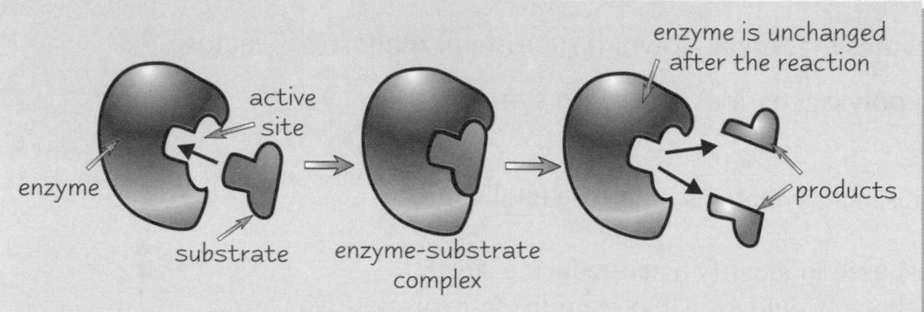


Scientists soon realised that the lock and key model didn't give the full story. The enzyme and substrate do have to fit together in the first place, but new evidence showed that the **enzyme-substrate complex changed shape** slightly to complete the fit. This **locks** the substrate even more tightly to the enzyme. Scientists modified the old lock and key model and came up with the '**induced fit**' model.

# Enzyme Action

## ...but the 'Induced Fit' Model is a Better Theory

The **'induced fit'** model helps to explain why enzymes are so **specific** and only bond to one particular substrate. The substrate doesn't only have to be the right shape to fit the active site, it has to make the active site **change shape** in the right way as well. This is a prime example of how a widely accepted theory can **change** when **new evidence** comes along. The 'induced fit' model is still widely accepted — for now, anyway.

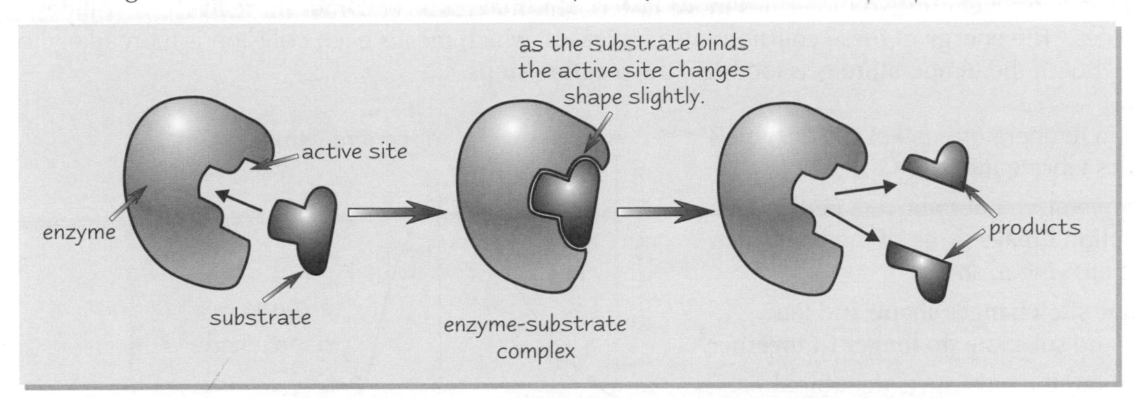

The 'Luminous Tights' model was popular in the 1980s but has since been found to be grossly inappropriate.

## Enzyme Properties relate to their Tertiary Structure

1) Enzymes are **very specific** — they usually only catalyse **one** reaction, e.g. maltase only breaks down maltose, sucrase only breaks down sucrose.
2) This is because **only one substrate will fit** into the active site.
3) The active site's **shape** is determined by the enzyme's **tertiary structure** (which is determined by the enzyme's **primary structure**).
4) Each **different enzyme** has a **different tertiary structure** and so a **different shaped active site**. If the substrate shape doesn't match the active site, the reaction won't be catalysed.
5) If the tertiary structure of a protein is **altered** in any way, the **shape** of the active site will **change**. This means the **substrate won't fit** into the active site and the enzyme will no longer be able to carry out its function.
6) The tertiary structure of an enzyme may be **altered** by changes in **pH** or **temperature** (see next page).
7) The **primary structure** (amino acid sequence) of a protein is determined by a **gene**. If a mutation occurs in that gene (see p. 53), it could change the tertiary structure of the enzyme **produced**.

## Practice Questions

Q1 What is an enzyme?

Q2 What is the name given to the amount of energy needed to start a reaction?

Q3 What is an enzyme-substrate complex?

Q4 Why can an enzyme only bind one substance?

**Exam Questions**

Q1 Describe the 'lock and key' model of enzyme action and explain how the 'induced fit' model is different. [7 marks]

Q2 Explain how a change in the amino acid sequence of an enzyme may prevent it from functioning properly. [2 marks]

## But why is the enzyme-substrate complex?

*So enzymes lower the activation energy of a reaction. I like to think of it as an assault course (bear with me). Suppose the assault course starts with a massive wall — enzymes are like the person who gives you a leg up over the w (see?). Without it you'd need lots of energy to get over the wall yourself and complete the rest of the course. Unli*

# Factors Affecting Enzyme Activity

*Well, here we are again... more about enzymes. You can't just bung an enzyme into a reaction and expect it to work. They're temperamental things, bless 'em, and require special conditions...*

## ***Temperature*** *has a* ***Big Influence*** *on Enzyme Activity*

Like any chemical reaction, the **rate** of an enzyme-controlled reaction **increases** when the **temperature's increased.** More heat means **more kinetic energy**, so molecules **move faster**. This makes the enzymes **more likely** to **collide** with the substrate molecules. The **energy** of these collisions also **increases**, which means each collision is more likely to **result** in a **reaction**. But, if the temperature gets too high, the **reaction stops**.

1) The rise in temperature makes the enzyme's molecules **vibrate more**.
2) If the temperature goes above a certain level, this vibration **breaks** some of the **bonds** that hold the enzyme in shape.
3) The **active site changes shape** and the enzyme and substrate **no longer fit together**.
4) At this point, the enzyme is **denatured** — it no longer functions as a catalyst.

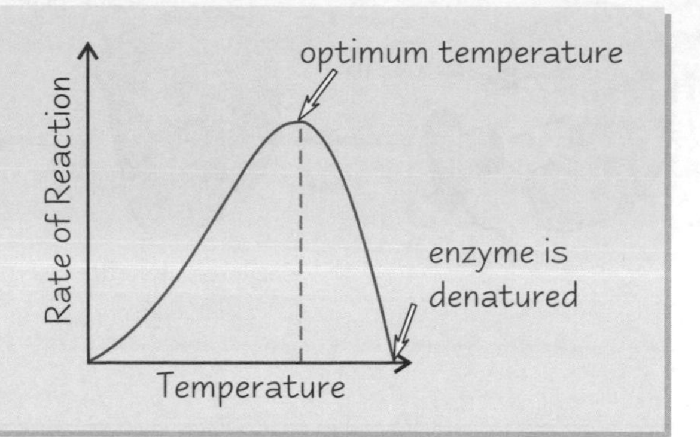


Every enzyme has an optimum temperature. For most human enzymes it's around 37 °C, but some enzymes, like those used in biological washing powders, can work well at 60 °C.

## ***pH*** *Also Affects Enzyme Activity*

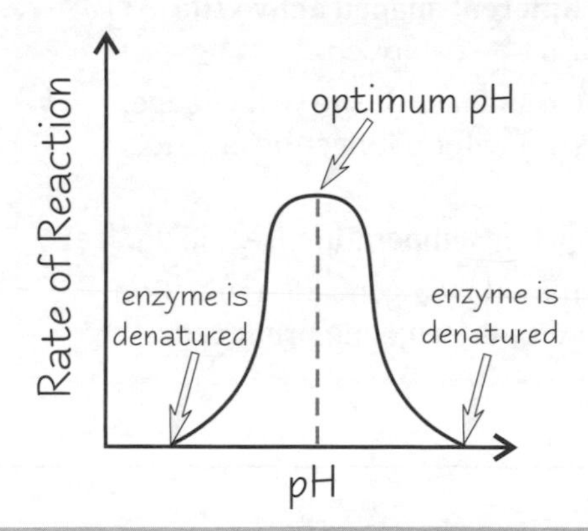


All enzymes have an **optimum pH value**. Most human enzymes work best at pH 7 (neutral), but there are exceptions. **Pepsin**, for example, works best at pH 2 (acidic), which is useful because it's found in the stomach. Above and below the optimum pH, the $H^+$ and $OH^-$ ions found in acids and alkalis can mess up the **ionic bonds** and **hydrogen bonds** that hold the enzyme's tertiary structure in place. This makes the **active site change shape**, so the enzyme is **denatured**.

## ***Substrate Concentration*** *Affects the Rate of Reaction* ***Up to a Point***

The **higher** the substrate concentration, the **faster** the reaction — more substrate molecules means a **collision** between substrate and enzyme is **more likely** and so more active sites will be used.
This is only true up until a '**saturation' point** though. After that, there are so many substrate molecules that the enzymes have about as much as they can cope with (all the **active sites are full**), and adding more **makes no difference**.

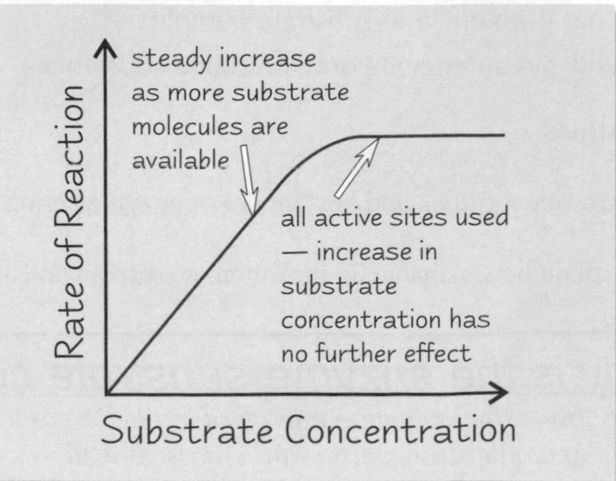

# Factors Affecting Enzyme Activity

## Enzyme Activity can be ***Inhibited***

Enzyme activity can be prevented by **enzyme inhibitors** — molecules that **bind to the enzyme** that they inhibit. Inhibition can be **competitive** or **non-competitive**.

### COMPETITIVE INHIBITION

1) **Competitive inhibitor** molecules have a **similar shape** to that of **substrate** molecules.
2) They **compete** with the substrate molecules to **bind** to the **active site**, but **no reaction** takes place.
3) Instead they **block** the active site, so **no substrate** molecules can **fit** in it.
4) How much the enzyme is inhibited depends on the **relative concentrations** of the inhibitor and substrate.
5) If there's a **high concentration** of the **inhibitor**, it'll take up nearly **all** the **active sites** and hardly any of the substrate will get to the enzyme.

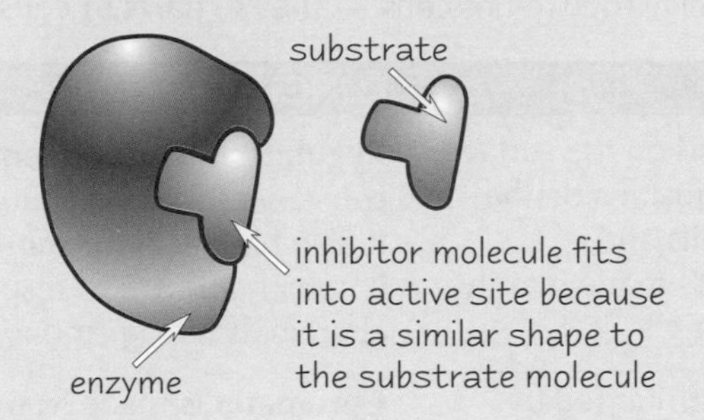


### NON-COMPETITIVE INHIBITION

1) **Non-competitive inhibitor** molecules bind to the enzyme **away from its active site.**
2) This causes the active site to **change shape** so the substrate molecules can no longer bind to it.
3) They **don't** 'compete' with the substrate molecules to bind to the active site because they are a **different shape.**
4) **Increasing** the concentration of **substrate won't** make any difference — enzyme activity will still be inhibited.

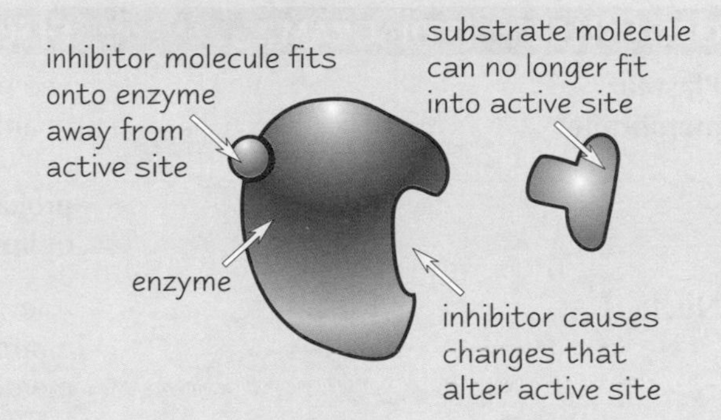


## Practice Questions

Q1 Draw a graph to show the effect of temperature on enzyme activity.

Q2 Draw a graph to show the effect of pH on enzyme activity.

Q3 Explain the effect of increasing substrate concentration on the rate of an enzyme-catalysed reaction.

**Exam Questions**

Q1 When doing an experiment on enzymes, explain why it is necessary to control the temperature and pH of the solutions involved. [8 marks]

Q2 Inhibitors prevent enzymes from working properly. They can be competitive or non-competitive.

a) Explain how a competitive inhibitor works. [3 marks]

b) Explain how a non-competitive inhibitor works. [2 marks]

## Activity — mine is usually inhibited by pizza and a movie...

*Human enzymes work well under normal body conditions — a neutral pH and body temp of 37 °C. Many poisons are enzyme inhibitors, e.g. cyanide. Even though there are thousands of enzymes in our bodies, inhibiting just one of them can cause severe problems. Some drugs are enzyme inhibitors though, e.g. Viagra, aspirin and penicillin, so they're not all bad.*

# Animal Cell Structure

*You've probably taken one look at the big table on these pages and got the fear. Don't worry though — there's nothing too taxing here... And besides, who doesn't love cells?*

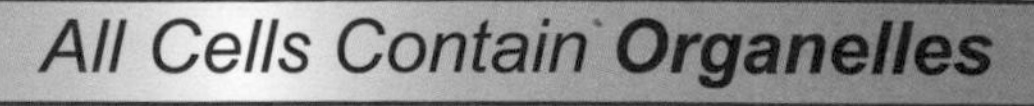

## All Cells Contain Organelles

Organelles are **parts of** cells. Each one has a **specific function**.

If you examine a cell through an **electron microscope** (see p. 24) you can see its **organelles** and the **internal structure** of most of them. Most of what's known about cell structure has been discovered by electron microscope studies. The diagram on the right shows the major parts of an **animal cell**.

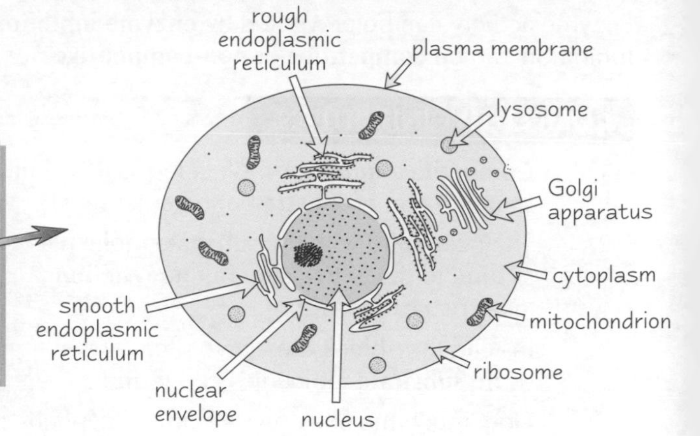


## Different Organelles have Different Functions

This giant table contains a big list of organelles — you need to know the **structure** and **function** of them all. Sorry. Most organelles are surrounded by **membranes**, which sometimes causes confusion — don't make the mistake of thinking that a diagram of an organelle is a diagram of a whole cell. They're not cells — they're **parts of** cells.

| ORGANELLE | DIAGRAM | DESCRIPTION | FUNCTION |
|---|---|---|---|
| **Plasma membrane** | | The membrane found on the surface of **animal cells** and just inside the cell wall of **plant cells** and **prokaryotic cells**. It's made mainly of **lipids** and **protein**. | **Regulates the movement** of substances into and out of the cell. It also has **receptor molecules** on it, which allow it to respond to chemicals like hormones. |
| **Nucleus** | | A large organelle surrounded by a **nuclear envelope** (double membrane), which contains many **pores**. The nucleus contains **chromatin** and often a structure called the **nucleolus**. | **Chromatin** is made from proteins and DNA (DNA **controls the cell's activities**). The pores allow substances (e.g. RNA) to move between the nucleus and the cytoplasm. The **nucleolus** makes **ribosomes** (see below). |
| **Lysosome** | | A **round organelle** surrounded by a **membrane**, with no clear internal structure. | Contains **digestive enzymes**. These are kept separate from the cytoplasm by the surrounding membrane, and can be used to **digest invading cells** or to **break down** worn out components of the cell. |
| **Ribosome** | | A **very small organelle** that floats free in the cytoplasm or is attached to the rough endoplasmic reticulum. | The **site** where **proteins** are **made**. |
| **Endoplasmic Reticulum (ER)** | | There are two types of endoplasmic reticulum: the **smooth endoplasmic reticulum** (diagram **a**) is a system of membranes enclosing a fluid-filled space. The **rough endoplasmic reticulum** (diagram **b**) is similar, but is **covered in ribosomes**. | The **smooth endoplasmic reticulum synthesises** and **processes lipids**. The **rough endoplasmic reticulum folds** and **processes proteins** that have been made at the ribosomes. |

# Animal Cell Structure

| ORGANELLE | DIAGRAM | DESCRIPTION | FUNCTION |
|---|---|---|---|
| Golgi Apparatus | | A group of fluid-filled **flattened sacs.** | It **processes** and **packages** new lipids and proteins. It also **makes lysosomes.** |
| Microvilli | microvilli, plasma membrane | These are **folds** in the plasma membrane. | They're found on cells involved in processes like absorption, such as epithelial cells in the small intestine (see p. 12). They **increase** the **surface area** of the plasma membrane. |
| Mitochondrion | outer membrane, inner membrane, crista, matrix | They're usually oval-shaped. They have a **double membrane** — the inner one is folded to form structures called **cristae**. Inside is the **matrix**, which contains enzymes involved in respiration. | The **site of aerobic respiration**. They're found in large numbers in cells that are very **active** and require a lot of **energy**. |

## Cells Have *Different Organelles* Depending on Their *Function*

In the exam, you might get a question where you need to apply your knowledge of the **organelles** in a cell to explain why it's particularly **suited** to its **function**. Here are some tips:

- Think about how the **structure** of the cell might affect its **job** — e.g. if it's part of an **exchange surface** it might have organelles that **increase** the **surface area** (e.g. microvilli). If it **carries things** it might have **lost** some of its organelles to make **more room**.
- Think about **what** the cell **needs** to do its **job** — e.g. if the cell uses a lot of **energy**, it'll need lots of **mitochondria**. If it makes a lot of **proteins** it'll need a lot of **ribosomes**.

You need to know the structure of an epithelial cell from the small intestine.

**Example** **Epithelial cells** in the **small intestine** are adapted to **absorb food efficiently.**

1) The walls of the small intestine have lots of finger-like projections called **villi** to **increase surface area.**
2) The **cells** on the surface of the villi have **microvilli** to increase surface area even more.
3) They also have **lots of mitochondria** — to provide **energy** for the transport of digested food molecules into the cell.

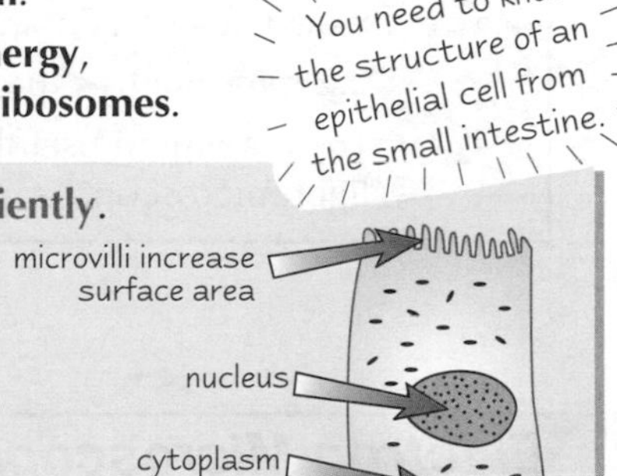


## Practice Questions

Q1 What is the function of a ribosome?

Q2 What is the function of a mitochondrion?

**Exam Questions**

Q1 Pancreatic cells make and secrete hormones (made of protein) into the blood. From production to secretion, list the organelles involved in making hormones. [4 marks]

Q2 Cilia are hair-like structures found on lung epithelial cells. Their function is to beat and move mucus out of the lungs. Beating requires energy. Suggest how ciliated cells are adapted to their function in terms of the organelles they contain. Explain your answer. [2 marks]

## The function of an organelle — to play music...

*Not the most exciting pages in the world but you need to know what all the organelles listed do. I'm afraid they'll keep popping up throughout the rest of the book — microvilli are important in digestion, the plasma membrane is essential for controlling the movement of things in and out of the cell and all the DNA stuff happens in the nucleus.*

# Analysis of Cell Components

*If you were born over a century ago then you wouldn't have had to learn all this stuff about organelles because people wouldn't have known anything about them. But then better microscopes were invented and here we are.*

## ***Magnification** is **Size**, **Resolution** is **Detail***

We all know that microscopes produce a **magnified image** of a sample, but **resolution** is just as important...

1) **MAGNIFICATION** is how much **bigger** the image is than the specimen (the sample you're looking at). It's calculated using this formula:

$$\text{magnification} = \frac{\text{length of image}}{\text{length of specimen}}$$

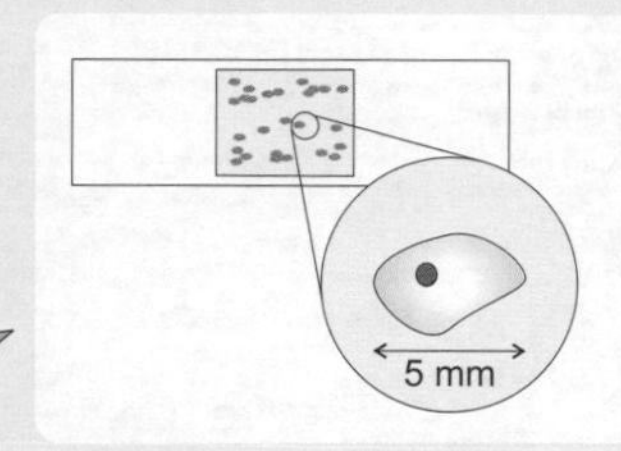


**For example:**
If you have a magnified image that's 5 mm wide and your specimen is 0.05 mm wide the magnification is:
5 ÷ 0.05 = **× 100**.

2) **RESOLUTION** is how **detailed** the image is. More specifically, it's how well a microscope **distinguishes** between **two points** that are **close together**. If a microscope lens can't separate two objects, then increasing the magnification won't help.

## *There are **Two Main Types** of Microscope — **Light** and **Electron***

**Light microscopes**

1) They use **light** (no surprises there).
2) They have a **lower resolution** than electron microscopes.
3) They have a maximum resolution of about **0.2 micrometres** (μm).
4) The maximum useful **magnification** of a light microscope is about **× 1500**.

**Electron microscopes**

1) They use **electrons** instead of light to form an image.
2) They have a **higher resolution** than light microscopes so give a **more detailed image**.
3) They have a maximum resolution of about **0.0001 micrometres** (μm). (About 2000 times higher than light microscopes.)
4) The maximum useful **magnification** of an electron microscope is about **× 1 500 000**.

A micrometre (μm) is one millionth of a metre, or 0.001 mm.

## ***Electron Microscopes** are either **'Scanning'** or **'Transmission'***

There are **two** types of **electron microscope**:

**Transmission electron microscopes (TEMs)**

1) TEMs use **electromagnets** to focus a **beam of electrons**, which is then transmitted **through** the specimen.
2) **Denser** parts of the specimen absorb **more electrons**, which makes them look **darker** on the image you end up with.
3) TEMs are good because they give **high resolution images**.
4) But they can only be used on **thin specimens**.

Nancy's New Year resolution was to get a hair cut.

**Scanning electron microscopes (SEMs)**

1) SEMs **scan** a beam of electrons across the specimen.
2) This **knocks off** electrons from the **specimen**, which are gathered in a **cathode ray tube** to form an **image**.
3) The images you end up with show the **surface** of the specimen and they can be **3-D**.
4) SEMs are good because they can be used on **thick specimens**.
5) But they give **lower resolution images** than TEMs.

# Analysis of Cell Components

## Cell Fractionation Separates Organelles

Suppose you wanted to look at some **organelles** under an **electron microscope**. First you'd need to **separate** them from the **rest of the cell** — you can do this by **cell fractionation**. There are **three** steps to this technique:

### 1 Homogenisation — Breaking Up the Cells

Homogenisation can be done in several **different ways**, e.g. by vibrating the cells or by grinding the cells up in a blender. This **breaks up** the **plasma membrane** and **releases** the **organelles** into solution.

### 2 Filtration — Getting Rid of the Big Bits

Next, the homogenised cell solution is **filtered** through a **gauze** to separate any **large cell debris** or **tissue debris**, like connective tissue, from the organelles. The organelles are much **smaller** than the debris, so they pass through the gauze.

### 3 Ultracentrifugation — Separating the Organelles

After filtration, you're left with a solution containing a **mixture** of organelles. To separate a particular organelle from all the others you use **ultracentrifugation**.

1) The cell fragments are poured into a **tube**. The tube is put into a **centrifuge** (a machine that separates material by spinning) and is spun at a **low speed**. The **heaviest organelles**, like nuclei, get flung to the **bottom** of the tube by the centrifuge. They form a **thick sediment** at the bottom — the **pellet**. The rest of the organelles stay suspended in the fluid above the sediment — the **supernatant**.
2) The supernatant is **drained off**, poured into **another tube**, and spun in the centrifuge at a **higher speed**. Again, the **heaviest organelles**, this time the mitochondria, form a pellet at the bottom of the tube. The supernatant containing the rest of the organelles is drained off and spun in the centrifuge at an **even higher speed**.
3) This process is **repeated** at higher and higher speeds, until all the organelles are **separated out**. Each time, the pellet at the bottom of the tube is made up of lighter and lighter organelles.

As the ride got faster, everyone felt their nuclei sink to their toes...

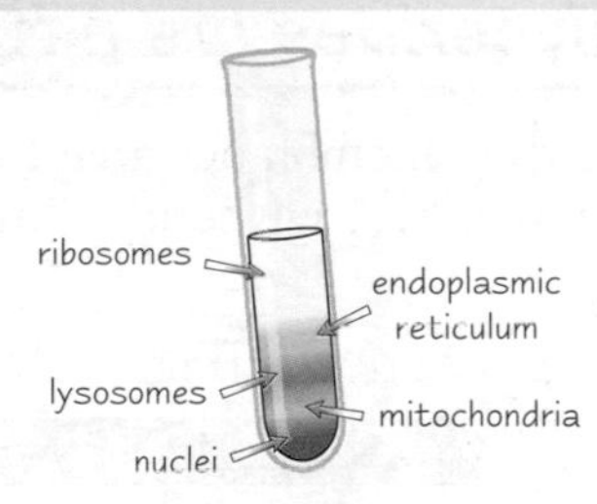


The organelles are separated in order of mass (from heaviest to lightest) — this order is usually: nuclei, then mitochondria, then lysosomes, then endoplasmic reticulum, and finally ribosomes.

## Practice Questions

Q1 What is meant by a microscope's magnification?

Q2 What is meant by a microscope's resolution?

**Exam Questions**

Q1 Describe the difference between SEMs and TEMs and give one limitation of each. [6 marks]

Q2 Describe how you would separate organelles from a cell sample using cell fractionation. Explain why each step is done. [8 marks]

## Cell fractionation — sounds more like maths to me...

*So, if you fancy getting up close and personal with mitochondria remember to homogenise, filter and ultracentrifuge first. Easy peasy. Then you have to decide if you want to use an SEM or TEM to view them, taking into account each of their limitations. Finally, you get to look at pretty pictures of what's inside our cells and make sounds like 'oooh' or 'ahhh'...*

# Plasma Membranes

*The plasma membrane is basically the cell boundary. To understand how substances get across this boundary (so they can enter or leave the cell) you have to know what it's made of. All I can say is... this section does get better.*

## Substances are Exchanged Across Plasma Membranes

In order to survive and carry out their functions, cells need to **take in** substances like glucose and oxygen, and **get rid of** substances like urea and carbon dioxide. The **plasma membrane** is a complex structure that controls what substances **enter** or **leave** the cell.

## Plasma Membranes are Mostly Made of Lipids

The **structure** of all **membranes** is basically the same. They're composed of **lipids** (mainly **phospholipids**), **proteins** and **carbohydrates** (usually attached to proteins or lipids).

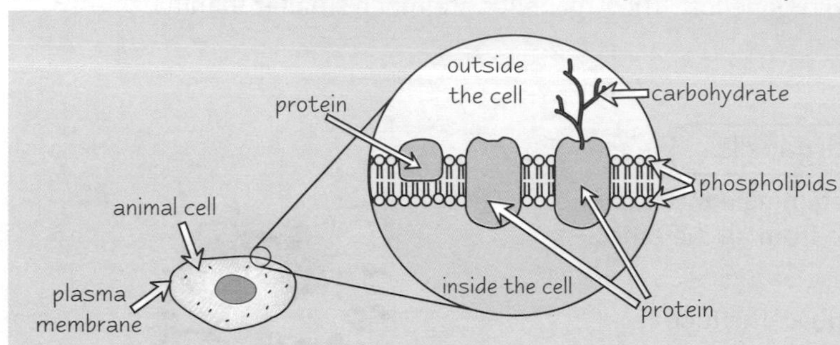


In 1972, the **fluid mosaic model** was suggested to describe the arrangement of molecules in the membrane. In the model, **phospholipid molecules** form a continuous, double layer (**bilayer**). This layer is 'fluid' because the phospholipids are constantly moving. **Protein molecules** are scattered through the layer, like tiles in a **mosaic**.

## Triglycerides are a Kind of Lipid

The lipids in membranes **aren't** triglycerides, but you need to know about triglycerides before you can understand phospholipids. Triglycerides have **one** molecule of **glycerol** with **three fatty acids** attached to it.

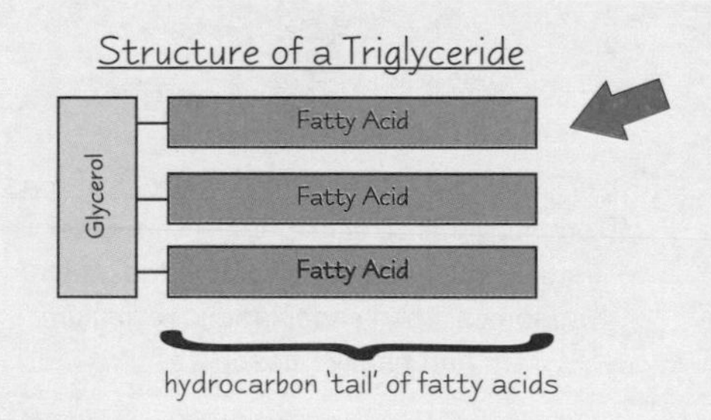


Fatty acid molecules have long 'tails' made of **hydrocarbons**. The tails are '**hydrophobic**' (they repel water molecules). These tails make lipids insoluble in water. All **fatty acids** have the same basic structure, but the **hydrocarbon tail varies.**

Basic Structure of a Fatty Acid

carbon atom links fatty acid to glycerol

O=C(OH)—R

variable 'R' group hydrocarbon tail

## Triglycerides are Formed by Condensation Reactions

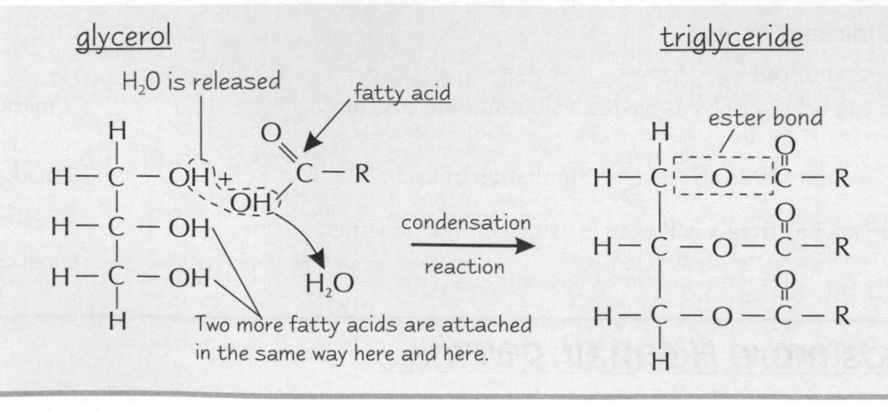


The diagram shows a **fatty acid** joining to a **glycerol molecule.** When the **ester bond** is formed a molecule of **water** is **released.** — it's a **condensation reaction**. This process happens twice more to form a **triglyceride**.

# Plasma Membranes

## Fatty Acids can be Saturated or Unsaturated

There are **two** kinds of fatty acids — **saturated** and **unsaturated**. The difference is in their **hydrocarbon tails**.

**Saturated** fatty acids **don't** have any **double bonds** between their **carbon atoms**. The fatty acid is 'saturated' with hydrogen.

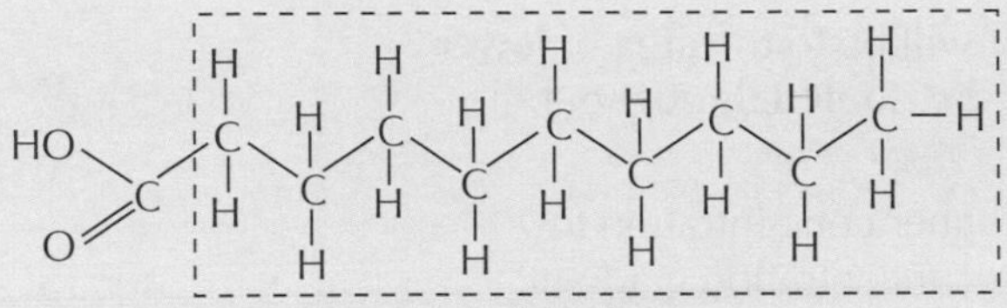

saturated hydrocarbon tail

**Unsaturated** fatty acids **do** have double bonds between **carbon atoms**, which cause the chain to kink.

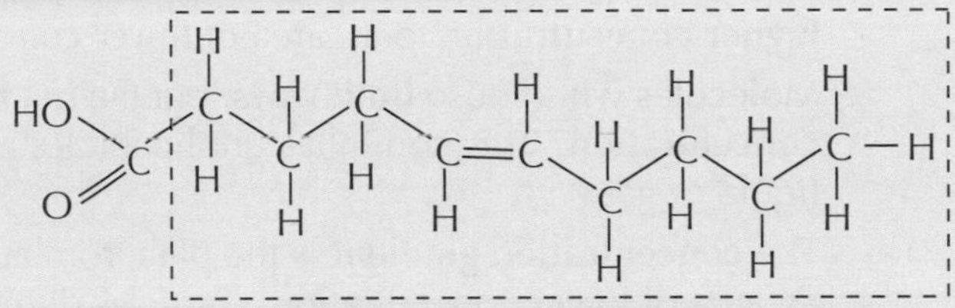

unsaturated hydrocarbon tail

## Phospholipids are Similar to Triglycerides

As you know, the lipids in **plasma membranes** are mainly **phospholipids**.

1) Phospholipids are pretty similar to triglycerides except one of the fatty acid molecules is replaced by a **phosphate group**.
2) The phosphate group is **hydrophilic** (attracts water). The fatty acid tails are **hydrophobic** (repel water). This is important in the plasma membrane (see p. 31 to find out why).

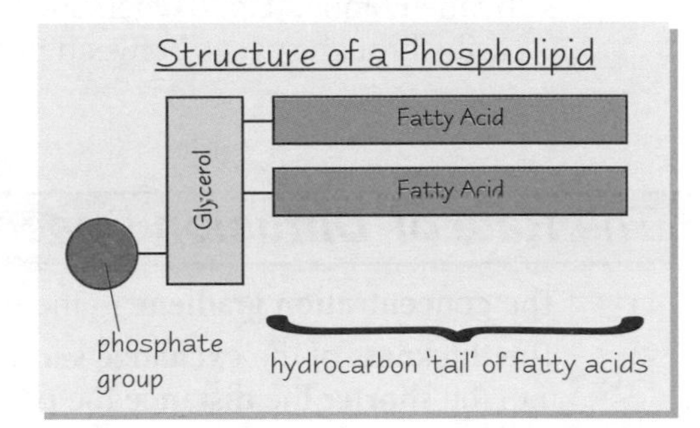


## Use the Emulsion Test for Lipids

If you wanted to find out if there was any **fat** in a particular **food** you could do the **emulsion test**:

1) **Shake** the test substance with **ethanol** for about a minute, then **pour** the solution into **water**.
2) Any lipid will show up as a **milky emulsion**.
3) The more lipid there is, the more noticeable the milky colour will be.

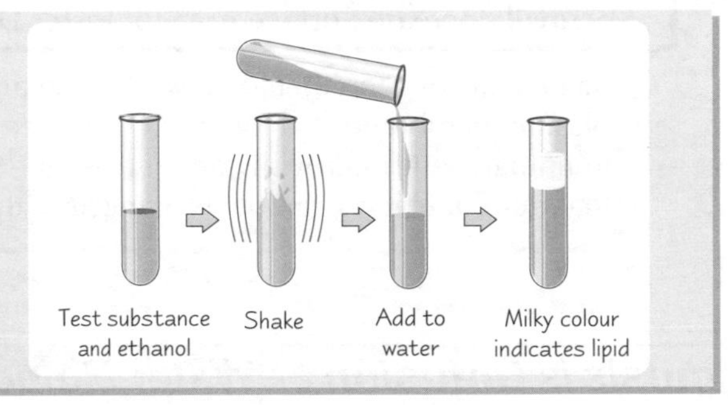


## Practice Questions

Q1 State three components of the plasma membrane.

Q2 Describe how you would test for lipids in a solution.

**Exam Question**

Q1 Explain why the plasma membrane can be described as having a fluid-mosaic structure. [2 marks]

Q2 Plasma membranes contain phospholipids.

a) Describe the structure of a phospholipid. [3 marks]

b) Explain the difference between a saturated fatty acid and an unsaturated fatty acid. [2 marks]

## The test for lipids — stick them in a can of paint...

*Not really. Otherwise you might upset your Biology teacher a bit. Instead, why not sit and contemplate all those phospholipids jumping around in your plasma membranes... their water-loving, phosphate heads poking out of the cell and into the cytoplasm, and their water-hating, hydrocarbon tails forming an impenetrable layer in between...*

# Exchange Across Plasma Membranes

*Ooooh it's starting to get a bit more exciting... here's how some substances can get across the plasma membrane without using energy. Just what you've always wanted to know, I bet.*

## **Diffusion** is the **Passive Movement** of **Particles**

1) Diffusion is the net movement of particles (molecules or ions) from an area of **higher concentration** to an area of **lower concentration.**
2) Molecules will diffuse **both ways**, but the **net movement** will be to the area of **lower concentration**. This continues until particles are **evenly distributed** throughout the liquid or gas.
3) The **concentration gradient** is the path from an area of higher concentration to an area of lower concentration. Particles diffuse **down** a concentration gradient.
4) Diffusion is a **passive process** — **no energy** is needed for it to happen.
5) Particles can diffuse **across plasma membranes**, as long as they can **move freely** through the membrane. E.g. oxygen and carbon dioxide molecules are **small enough** to pass easily through spaces between phospholipids.

## The **Rate of Diffusion** Depends on **Several Factors**

1) The **concentration gradient** — the **higher** it is, the **faster** the rate of diffusion.
2) The **thickness** of the **exchange surface** — the **thinner** the exchange surface (i.e. the **shorter** the **distance** the particles have to travel), the **faster** the rate of diffusion.
3) The **surface area** — the **larger** the surface area (e.g. of the plasma membrane), the **faster** the rate of diffusion.

### **Microvilli** Increase **Surface Area** for **Faster Diffusion**

Some cells (e.g. epithelial cells in the small intestine) have **microvilli** — projections formed by the plasma membrane folding up on itself (see p. 23). Microvilli give the cell a **larger surface area** — in human cells microvilli can increase the surface area by about **600 times**. A larger surface area means that **more particles** can be **exchanged** in the same amount of time — **increasing the rate of diffusion.**

## **Osmosis** is **Diffusion** of **Water Molecules**

1) Osmosis is the **diffusion** of **water molecules** across a **partially permeable membrane**, from an area of **higher water potential** (i.e. higher concentration of water molecules) to an area of **lower water potential** (i.e. lower concentration of water molecules).
2) **Water potential** is the potential (likelihood) of water molecules to diffuse out of or into a solution.
3) A **partially permeable membrane** allows some molecules through it, but not all.
4) The **plasma membrane** is **partially permeable.** Water molecules are small and can diffuse easily through the **plasma membrane**, but large solute molecules can't.
5) **Pure water** has the **highest water potential.** All solutions have a **lower** water potential than pure water.

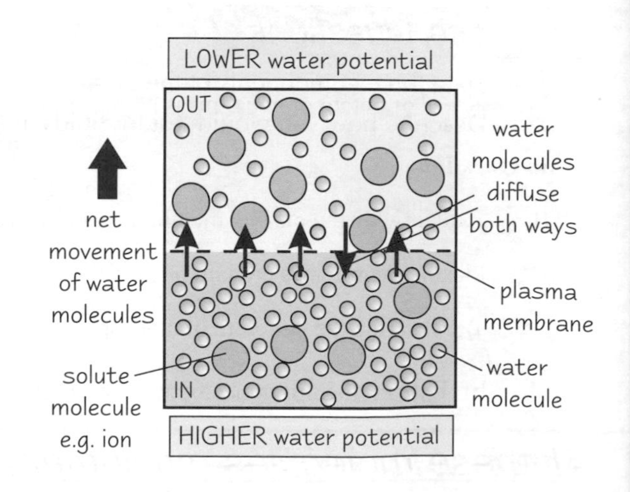

# Exchange Across Plasma Membranes

## Facilitated Diffusion uses Carrier Proteins and Protein Channels

1) Some **larger molecules** (e.g. amino acids, glucose) and **charged atoms** (e.g. chloride ions) **can't diffuse directly through** the phospholipid bilayer of the cell membrane.
2) Instead they diffuse through **carrier proteins** or **protein channels** in the cell membrane — this is called **facilitated diffusion.**
3) Like diffusion, facilitated diffusion moves particles **down** a **concentration gradient**, from a higher to a lower concentration.
4) It's also a passive process — it **doesn't** use **energy**.

Andy needed all his concentration for this particular gradient...

**Carrier proteins** move **large molecules** into or out of the cell, down their concentration gradient. **Different carrier proteins** facilitate the diffusion of **different molecules.**

1) First, a large molecule **attaches** to a carrier protein in the membrane.
2) Then, the protein **changes shape.**
3) This **releases** the molecule on the **opposite side** of the membrane.

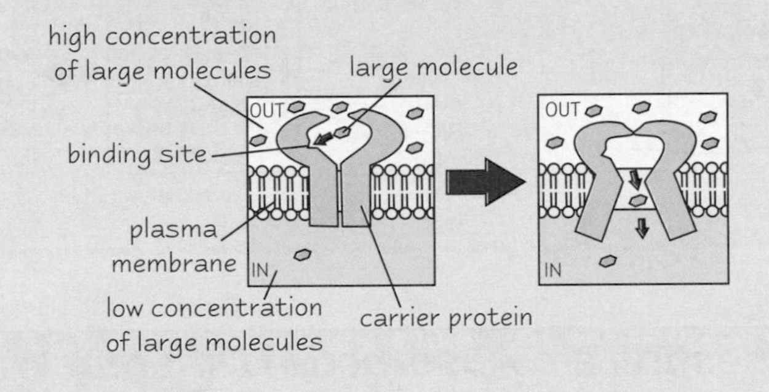


**Protein channels** form **pores** in the membrane for **charged particles** to diffuse through (down their concentration gradient). **Different protein channels** facilitate the diffusion of **different charged particles.**

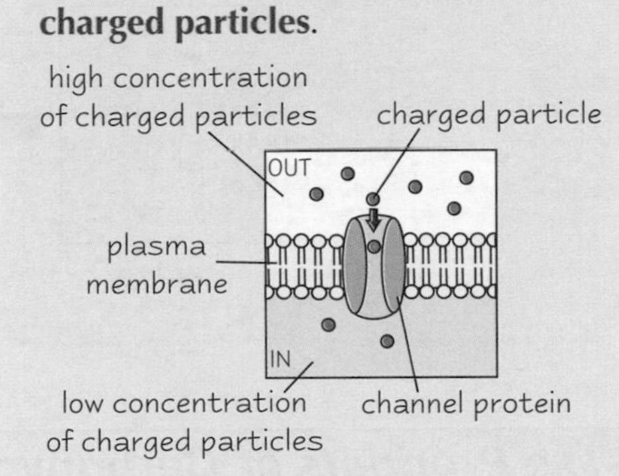


## Practice Questions

Q1 Diffusion is a passive process. What does this mean?

Q2 Give two factors that affect the rate of diffusion.

Q3 How do microvilli increase the rate of diffusion?

Q4 What is facilitated diffusion?

**Exam Question**

Q1 Pieces of potato of equal mass were put into different concentrations of sucrose solution for three days. The difference in mass for each is recorded in the table on the right.

| Concentration of sucrose / % | 1 | 2 | 3 | 4 |
|---|---|---|---|---|
| Mass difference / g | 0.4 | 0.2 | 0 | – 0.2 |

a) Explain why the pieces of potato in 1% and 2% sucrose solutions gained mass. [3 marks]

b) Suggest a reason why the mass of the piece of potato in 3% sucrose solution stayed the same. [1 mark]

c) What would you expect the mass difference for a potato in a 5% solution to be? Explain your answer. [4 marks]

## All these molecules moving about — you'd think they'd get tired...

*Right, I think I get it. If you're a small molecule, like oxygen, you can just cross the membrane by simple diffusion. If you're a water molecule you can also cross the membrane by diffusion, but you call it a fancy name — osmosis. And if you're a large or charged molecule you have a little help from a channel or carrier protein. There's a transport process to suit everyone...*

# Exchange Across Plasma Membranes

*Diffusion and are osmosis are passive processes, so for those of you feeling a bit more active here's a page all about... you guessed it... active transport.*

## *Active Transport Moves Substances Against a Concentration Gradient*

Active transport uses **energy** to move **molecules** and **ions** across plasma membranes, **against** a **concentration gradient**.

**Carrier proteins** are also involved in active transport:

1) The process is pretty similar to facilitated diffusion — a molecule **attaches** to the carrier protein, the protein **changes shape** and this moves the molecule **across** the membrane, **releasing it** on the other side.
2) The only difference is that **energy** is used (from **ATP** — a common source of energy used in the cell), to move the solute against its concentration gradient.

The diagram shows the active transport of **calcium**.

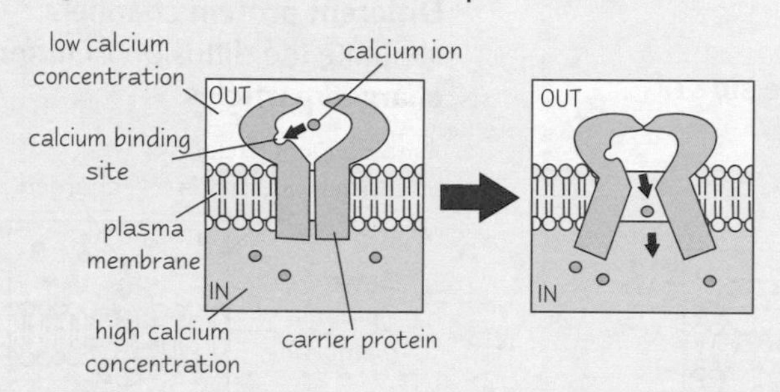


**Co-transporters** are a type of **carrier protein**.

1) They bind **two** molecules at a time.
2) The concentration gradient of one of the molecules is used to move the other molecule **against** its own concentration gradient.

The diagram shows the co-transport of **sodium ions** and **glucose**. Sodium ions move into the cell **down** their concentration gradient. This moves glucose into the cell too, **against** its concentration gradient.

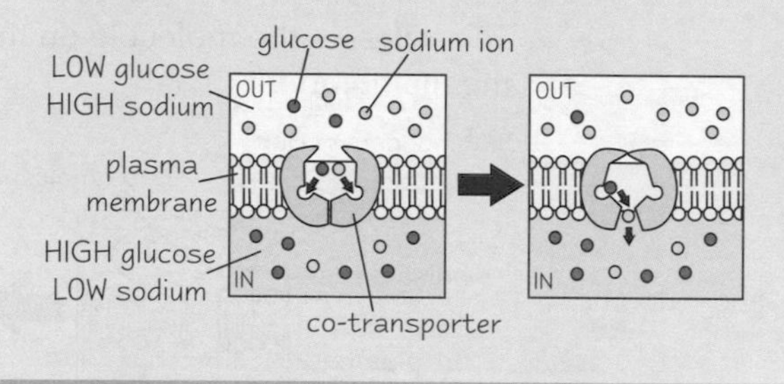


## *The Products of Carbohydrate Digestion are Absorbed in Different Ways*

All these processes — **diffusion**, **facilitated diffusion** and **active transport** — are **essential** in the body.
For example, to **absorb** the **products** of **carbohydrate digestion** (e.g. **glucose**) across the **intestinal epithelium cells**:

### Some **glucose diffuses** across the **intestinal epithelium** into the **blood**

When carbohydrates are first broken down, there's a **higher concentration** of **glucose** in the **small intestine** than in the **blood** — there's a **concentration gradient**. Glucose moves across the **epithelial cells** of the small intestine into the blood by **diffusion**. When the concentration in the lumen becomes **lower** than in the blood diffusion **stops**.

### Some **glucose** enters the **intestinal epithelium** by **active transport** with **sodium ions**

The remaining glucose is absorbed by **active transport**. Here's how it all works:

1) **Sodium ions** are **actively transported out** of the small intestine epithelial **cells**, into the **blood**, by the **sodium-potassium pump**. This creates a **concentration gradient** — there's now a higher concentration of sodium ions in the small intestine lumen than inside the cell.
2) This causes sodium ions to diffuse from the small intestine lumen into the cell, down their concentration gradient. They do this via the **sodium-glucose co-transporter proteins**.
3) The co-transporter carries **glucose** into the cell with the sodium. As a result the concentration of **glucose** inside the cell **increases**.
4) Glucose diffuses out of the cell, into the blood, down its concentration gradient through a protein channel, by **facilitated diffusion**.

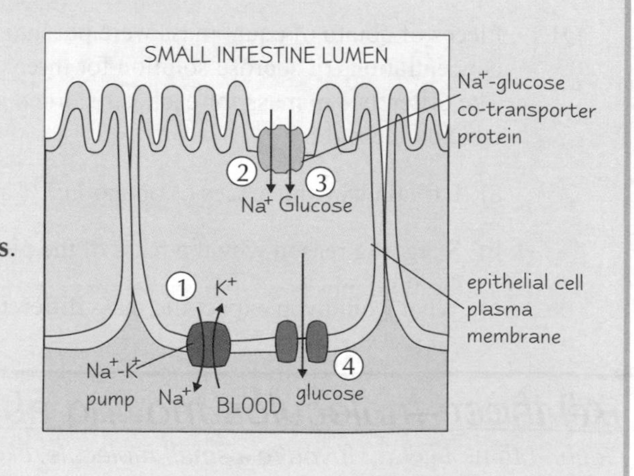

# Exchange Across Plasma Membranes

## You can use the **Fluid Mosaic Model** to Explain **Membrane Properties**

You might get a question in the exam where you need to use your **knowledge** of the **fluid mosaic model** to explain why the **plasma membrane** has various **properties**. You can't go far wrong if you learn these **five** points.

### 1) The membrane is a good barrier against most water-soluble molecules

**Phospholipids** are the major component of the membrane bilayer. The **hydrophobic tails** of the phospholipids make it difficult for **water-soluble** substances, such as sodium ions and glucose, to get through.

### 2) The membrane controls what enters and leaves

**Protein channels** and **carrier proteins** in the membrane allow the passage of large or charged **water-soluble** substances that would otherwise find it difficult to cross the membrane. **Different cells** have **different** protein channels and carrier proteins — e.g. the membrane of a **nerve cell** has many **sodium-potassium carrier proteins** (which help to conduct nerve impulses) and **muscle cells** have **calcium protein channels** (which are needed for muscle contraction).

### 3) The membrane allows cell communication

Membranes contain **receptor proteins.** These allow the cell to **detect chemicals** released from other cells. The chemicals **signal** to the cell to **respond** in some way, e.g. the hormone insulin binds to receptors in the membranes of liver cells — this tells the liver cells to absorb glucose. This cell communication is vital for the body to **function properly**. **Different cells** have **different receptors** present in their membranes.

### 4) The membrane allows cell recognition

Some **proteins** and **lipids** in the plasma membrane have short **carbohydrate chains** attached to them — they're called **glycoproteins** and **glycolipids**. These molecules tell **white blood cells** that the cell is **your own.** White blood cells only attack cells that they don't recognise as **self** (e.g. those of **microorganisms** like bacteria).

### 5) The membrane is fluid

The **phospholipids** in the plasma membrane are **constantly moving** around. The more **unsaturated** fatty acids there are in the phospholipid bilayer, the **more fluid** it becomes. **Cholesterol** molecules fit in between the phospholipids of the bilayer — the **more** cholesterol molecules there are, the **less fluid** the membrane becomes. Cholesterol is important as it makes the cell membrane more **rigid** and prevents it from **breaking up**.

## Practice Questions

Q1 What is active transport?

Q2 Describe how carrier proteins actively transport substances across the cell membrane.

Q3 Which molecule provides the energy for active transport?

Q4 Give three properties of membranes.

**Exam Questions**

Q1 Describe and explain how the glucose produced from starch digestion is absorbed into the blood by diffusion and active transport. [10 marks]

Q2 Use the fluid mosaic model of membrane structure to explain how the membrane controls what enters and leaves the cell. [5 marks]

## Revision — like working against a concentration gradient...

*Don't worry if it takes you a while to learn these pages — there's quite a lot to cover. It's a good idea to learn it bit by bit. Don't move on to co-transport until you fully understand active transport in normal carrier proteins. Then make sure you can explain each of the properties of the plasma membrane using the fluid mosaic model. They don't ask for much, do they...*

# Cholera

*It's not all cherries and pie though — exchange of substances in and out of cells can be disrupted by diseases like cholera...*

## The **Cholera Bacterium** is a **Prokaryotic** Organism

There are **two** types of organisms — **prokaryotic** and **eukaryotic**.
Prokaryotic organisms are made up of **prokaryotic cells** (i.e. they're single-celled organisms), and eukaryotic organisms are made up of **eukaryotic cells**.

1) Eukaryotic cells are **complex** and include all **animal** and **plant cells**.
2) Prokaryotic cells are **smaller** and **simpler**, e.g. bacteria.

You need to know the **structure** of a prokaryotic cell and what all the different organelles inside are for.

See pages 22-23 for more on organelles.

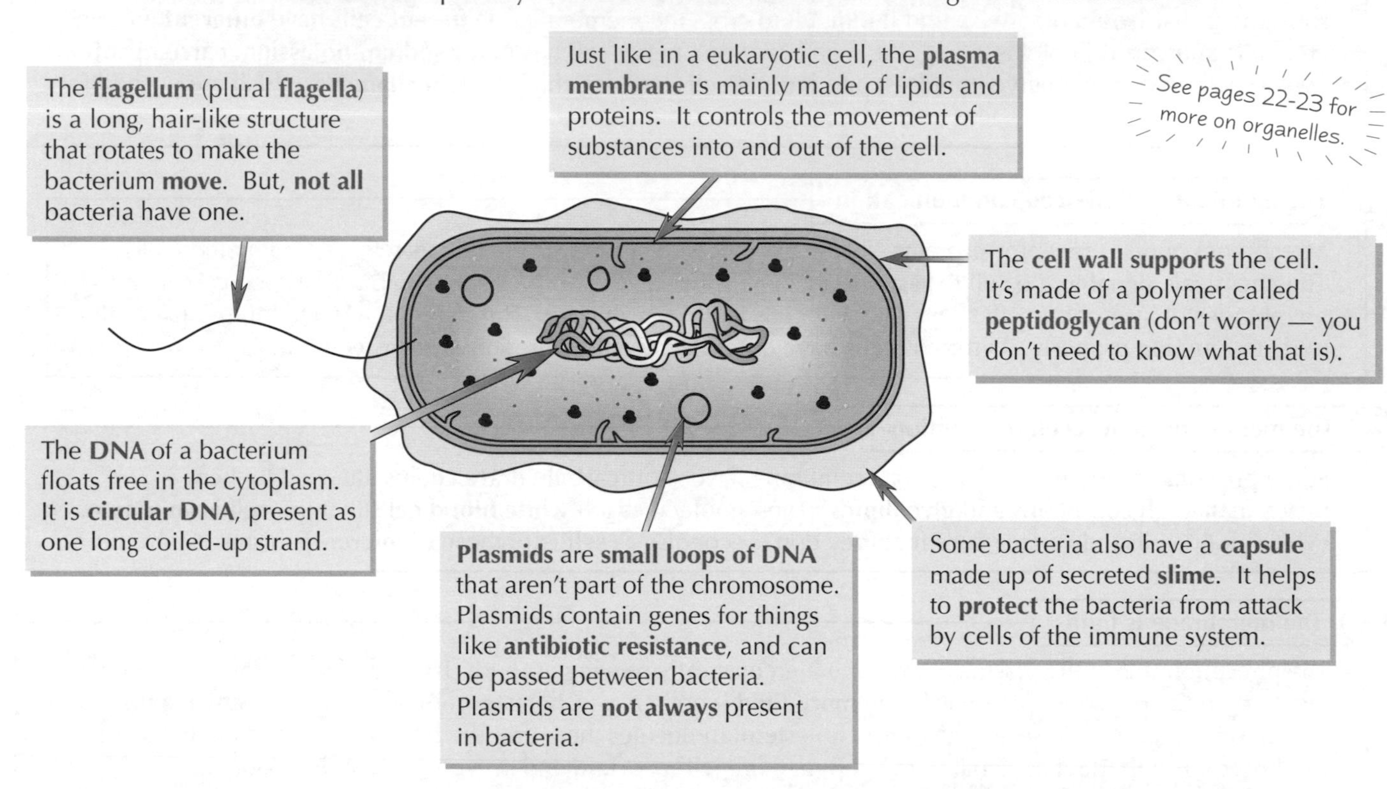


## **Cholera** Bacteria Produce a **Toxin** That Affects **Chloride Ion Exchange**

Cholera bacteria produce a **toxin** when they infect the body. This toxin causes a fair old bit of havoc...

1) The toxin causes **chloride ion protein channels** in the plasma membranes of the small intestine epithelial cells to **open**.
2) Chloride ions move **into the small intestine lumen**. The build up of chloride ions **lowers** the **water potential** of the lumen.
3) **Water** moves **out** of the **blood**, across the epithelial cells, and **into the small intestine lumen** by **osmosis** (to even up the water concentration).
4) The massive increase in water secretion into the intestine lumen leads to **really, really, really bad diarrhoea** — causing the body to become extremely **dehydrated**.

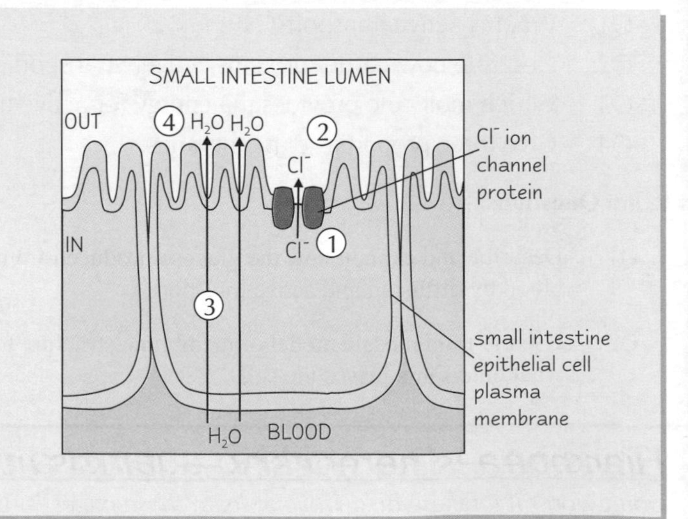

# Cholera

## **Oral Rehydration Solutions** are used to **Treat Diarrhoeal Diseases**

People suffering from **diarrhoeal diseases** like cholera need to **replace** all the **fluid** that they've **lost** in the diarrhoea. The quickest way to do this is by inserting a **drip** into a person's **vein**. However, not everywhere in the world has access to drips, so **oral rehydration solutions** are used instead.

**Oral Rehydration Solutions (ORSs)**

1) An oral rehydration solution is a **drink** that contains large amounts of **salts** (such as sodium ions and chloride ions) and **sugars** (such as glucose and sucrose) dissolved in water.
2) **Sodium** ions are included to increase **glucose** absorption (sodium and glucose are **co-transported** into the epithelium cells in the intestine — see p. 30).
3) Getting the **concentration** of the ORS right is essential for effective treatment.
4) An ORS is a very **cheap** treatment and the people administering it **don't** require much **training**. This makes it great for treating diarrhoeal diseases in **developing countries** (where they're a huge problem).

## New **Oral Rehydration Solutions** can be **Tested on Humans**

ORS are so important in treating diarrhoeal disease that research into the development of **new**, **improved** ORS is always being carried out. But before a new ORS can be put into use, scientists have to show that it's **more effective** than the old ORS and that it's **safe**. This is done by **clinical testing** on humans.

There are some **ethical issues** associated with **trialling ORS**

1) **Diarrhoeal diseases** mostly affect **children**, so many **trials** involve **children**. **Parents** decide whether the child will **participate** in the trial. The child doesn't make their **own decision** — some people think this is unethical.
2) But scientists believe the treatment must be trialled on children if it's to be shown to be **effective** against a **disease** that mainly affects children.
3) Clinical trials usually involve a **blind trial**. This is where some patients who are admitted into hospital with diarrhoeal diseases are given the **standard ORS** and others are given the **new ORS**. This means that the two can be **compared**. It's called a blind trial because the patients **don't know** which treatment they've been given. Some people don't agree with this — they think that people have the **right** to **know** and **decide** on the **treatment** that they're going to have.
4) Scientists argue that a blind trial is important to eliminate any **bias** that may **skew** the **data** as a result of **patients knowing** which treatment they've received.
5) When a new ORS is first trialled, there's no way of knowing whether it'll be **better** than the current ORS — there is a **risk** of the patient **dying** when the original, better treatment was available.

## Practice Questions

Q1 Give an example of a prokaryotic organism.

Q2 What is the function of a flagellum?

Q3 What are plasmids?

Q4 Suggest a reason why oral rehydration solutions contain sodium ions.

**Exam Question**

Q1 Explain how infection with the cholera bacterium leads to diarrhoea. [5 marks]

Q2 Give one argument for and one argument against trialling new oral rehydration solutions on children. [2 marks]

## Diarrhoea is hereditary — it runs in your jeans...

*Well, I don't know about you but all I can think about now is poo. With cholera, it's actually the poo that can kill you, so to speak, because you lose so much water with it. Antibiotics can help to clear the cholera bacterium, but if you aren't rehydrated straightaway you could die because your cells need plenty of water to carry out all their chemical reactions.*

# Lung Function

*To the examiners, lung function is more than just breathing in and out — they love it (and don't even get them started on gas exchange in the alveoli). So unsurprisingly it's a good idea if you know a bit about the lungs... take a deep breath...*

## *Lungs are Specialised Organs for Gas Exchange*

Humans need to get **oxygen** into the blood (for respiration) and they need to **get rid** of **carbon dioxide** (made by respiring cells). This is where **breathing** (or **ventilation** as it's sometimes called) and the **lungs** come in.

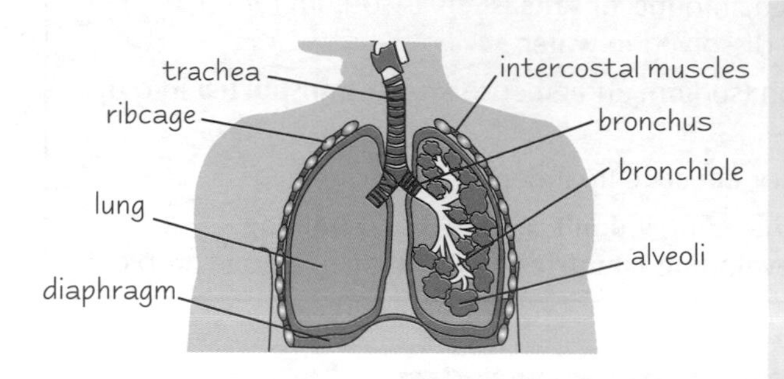


1) As you breathe in, air enters the **trachea** (windpipe).
2) The trachea splits into two **bronchi** — one **bronchus** leading to each lung.
3) Each bronchus then branches off into smaller tubes called **bronchioles**.
4) The bronchioles end in small 'air sacs' called **alveoli** (this is where gases are exchanged — see next page).
5) The **ribcage**, **intercostal muscles** and **diaphragm** all work together to move air in and out (see below).

## *Ventilation is Breathing In and Breathing Out*

**Ventilation** consists of **inspiration** (breathing in) and **expiration** (breathing out).

**Inspiration**

1) The **intercostal** and **diaphragm muscles contract**.
2) This causes the **ribcage to move upwards and outwards** and the **diaphragm to flatten**, **increasing the volume** of the thorax (the space where the lungs are).
3) As the volume of the thorax increases the lung pressure **decreases** (to below atmospheric pressure).
4) This causes air to flow **into the lungs**.
5) Inspiration is an **active process** — it requires **energy**.

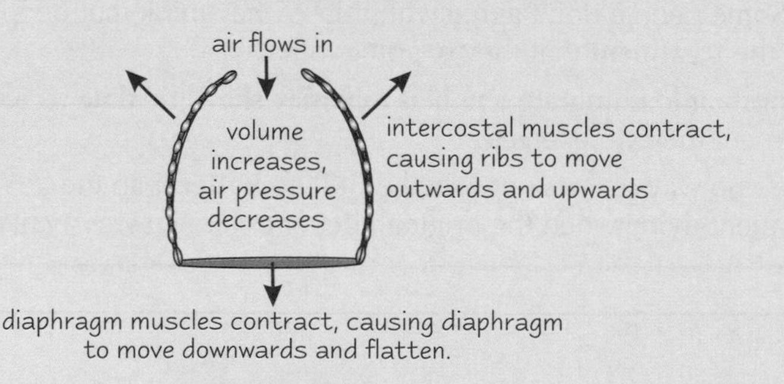

Thankfully, evolution came up with a better ventilation system.

**Expiration**

1) The **intercostal** and **diaphragm muscles relax**.
2) The **ribcage** moves **downwards and inwards** and the **diaphragm** becomes **curved** again.
3) The thorax volume **decreases**, causing the air pressure to **increase** (to above atmospheric pressure).
4) Air is forced **out of the lungs**.
5) Expiration is a **passive process** — it **doesn't** require energy.

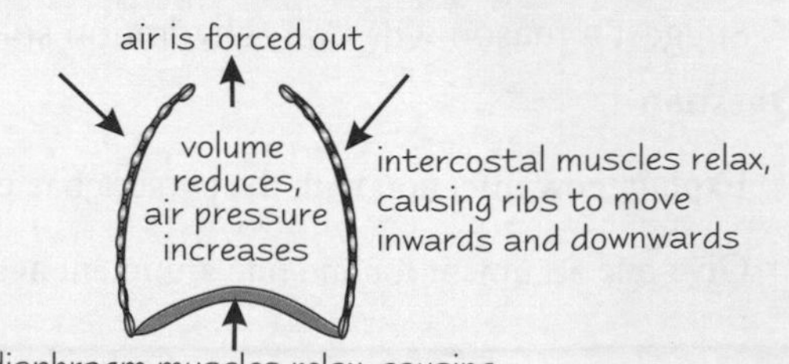

# Lung Function

## In Humans Gaseous Exchange Happens in the Alveoli

Lungs contain millions of microscopic air sacs where gas exchange occurs — called **alveoli**.
Each alveolus is made from a single layer of thin, flat cells called alveolar epithelium.

1) There's a huge number of alveoli in the lungs, which means there's a **big surface area** for exchanging oxygen ($O_2$) and carbon dioxide ($CO_2$).
2) The alveoli are surrounded by a network of **capillaries**.

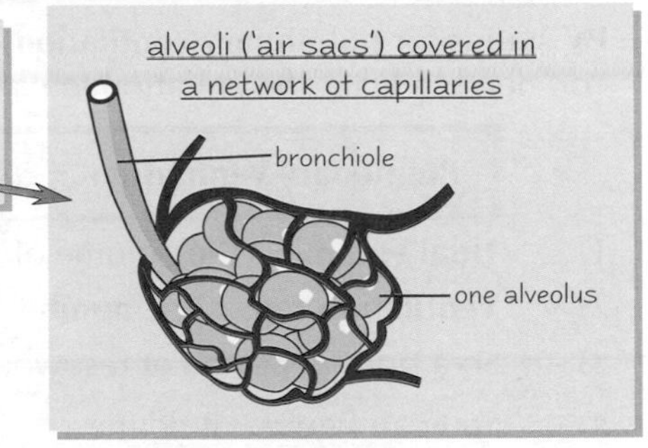


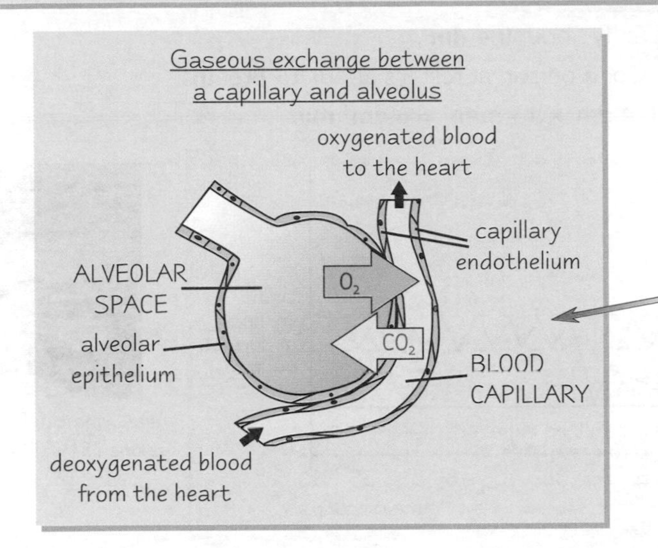


3) **$O_2$** diffuses **out of** the alveoli, across the **alveolar epithelium** and the **capillary endothelium** (a type of epithelium that forms the capillary wall), and into **haemoglobin** (see p. 58) in the **blood**.
4) **$CO_2$** diffuses **into** the alveoli from the blood, and is breathed out.

Epithelial tissue is pretty common in the body. It's usually found on exchange surfaces.

## The Alveoli are Adapted for Gas Exchange

Alveoli have features that **speed up** the **rate of diffusion** so gases can be exchanged quickly:

1) **A thin exchange surface** — the **alveolar epithelium** is only **one cell thick.** This means there's a **short diffusion pathway** (which speeds up diffusion).
2) **A large surface area** — the **large number** of alveoli means there's a large surface area for gas exchange.

There's also a **steep concentration gradient** of oxygen and carbon dioxide between the alveoli and the capillaries, which increases the rate of diffusion. This is constantly maintained by the **flow of blood** and **ventilation**.

## Practice Questions

Q1 Describe the structure of the human gas exchange system.

Q2 Describe the process of inspiration and expiration.

Q3 Describe the movement of carbon dioxide and oxygen across the alveolar epithelium.

**Exam Question**

Q1 Explain why there is a fast rate of gas exchange in the alveoli. [6 marks]

## Alveoli — useful things... always make me think about pasta...

*Just like the digestive system, a mammal's lungs act as an interface with the environment — they take in air and give out waste gases. Ventilation moves these gases into and out of the lungs, but the alveoli have the task of getting them in and out of the bloodstream. Luckily, like many other biological structures, they're well adapted for doing their job.*

# How Lung Disease Affects Function

*It's all very well when your lungs are working perfectly, but some pathogens (and even your lifestyle) can muck them up good and proper, reducing the rate of gas exchange. Not good.*

## PV is the Volume of Air Taken into the Lungs in One Minute

**PV** stands for **pulmonary ventilation** — it's measured in $dm^3\ min^{-1}$.
You need to learn the equation to calculate it:

**Pulmonary Ventilation = Tidal volume × Ventilation rate**

$dm^3\ min^{-1}$ is cubic decimetres per minute — a decimetre is 10 centimetres, and 1 $dm^3$ is the same as a litre.

1) **Tidal volume** is the volume of air in **each breath** — usually about **0.4 $dm^3$**.
2) **Ventilation rate** is the **number of breaths per minute**. For a person at rest it's about **15 breaths**.
3) So a normal person at rest would have a PV of about $0.4\ dm^3 \times 15\ min^{-1}$ = **$6\ dm^3\ min^{-1}$**.
4) You can figure out tidal volume and ventilation rate from the graph produced from a spirometer (a fancy machine that scientists and doctors use to measure the volume of air breathed in and out):

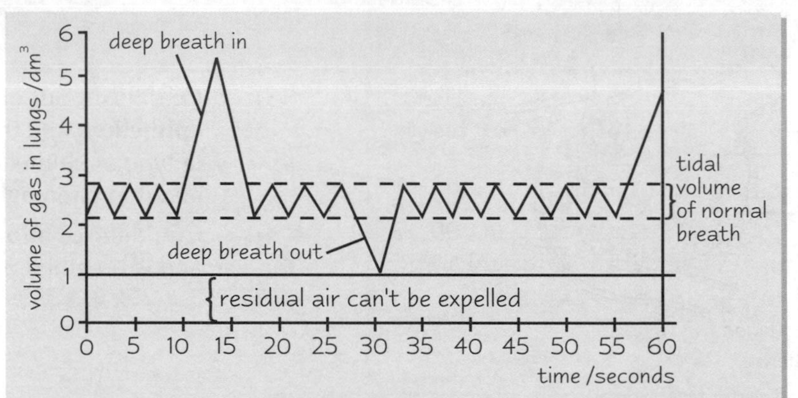

Measuring tidal volume is one of the hardest jobs in the world.

## Pulmonary Tuberculosis (TB) is a Lung Disease Caused by Bacteria

Pulmonary tuberculosis is caused by the bacterium ***Mycobacterium tuberculosis***.

**Infection**

1) When someone becomes infected with tuberculosis bacteria, immune system cells build a **wall** around the bacteria in the lungs. This forms small, hard lumps known as **tubercles**.
2) Infected tissue within the tubercles **dies**, the gaseous exchange surface is **damaged** so **tidal volume** is **decreased**.
3) Tuberculosis also causes **fibrosis** (see next page), which further reduces the tidal volume.
4) If the bacteria enter the **bloodstream**, they can **spread** to other parts of the body.

**Symptoms**

1) Common symptoms include a persistent **cough**, coughing up **blood** and **mucus**, **chest pains**, **shortness of breath** and **fatigue**.
2) Sufferers may also have a **fever**.
3) Many **lose weight** due to a reduction in appetite.

**Transmission**

1) TB is transmitted by **droplet infection** — when an infected person **coughs** or **sneezes**, tiny **droplets** of **saliva** and **mucus** containing the bacteria are released from their mouth and nose. If an uninfected person breathes in these droplets, the bacteria are **passed on**.
2) Tuberculosis tends to be much more widespread in areas where **hygiene** levels are **poor** and where people live in **crowded** conditions.
3) TB can be prevented with the BCG vaccine, and can be treated with antibiotics.

Many people with tuberculosis are **asymptomatic** — they're infected but they **don't show** any symptoms, because the infection is in an **inactive form**. People who are asymptomatic are unable to pass the infection on. But if they become **weakened**, for example by another disease or malnutrition, then the infection can become **active**. They'll show the symptoms and be able to pass on the infection.

# How Lung Disease Affects Function

## *Fibrosis, Asthma and Emphysema all Affect Lung Function*

Fibrosis, asthma and emphysema all **reduce the rate of gas exchange** in the alveoli. Less oxygen is able to diffuse into the bloodstream, the body cells **receive less oxygen** and the rate of **aerobic respiration** is **reduced**. This means **less energy is released** and sufferers often feel **tired** and **weak**.

### Fibrosis

1) Fibrosis is the formation of **scar tissue** in the lungs. This can be the result of an **infection** or exposure to substances like **asbestos** or **dust**.
2) Scar tissue is **thicker** and **less elastic** than normal lung tissue.
3) This means that the lungs are **less able to expand** and so **can't hold as much air** as normal — the tidal volume is **reduced**. It's also harder to **force air out** of the lungs due to the loss of elasticity.
4) There's a **reduction** in the rate of **gaseous exchange** — **diffusion** is **slower** across a **thicker** scarred membrane.
5) Symptoms of fibrosis include **shortness of breath**, a **dry cough**, **chest pain**, **fatigue** and **weakness**.
6) Fibrosis sufferers have a **faster breathing rate** than normal — to get enough air into their lungs to **oxygenate** their blood.

### Asthma

1) Asthma is a respiratory condition where the airways become **inflamed** and **irritated**. The causes vary from case to case but it's usually because of an **allergic reaction** to substances such as **pollen** and **dust**.
2) During an asthma attack, the **smooth muscle** lining the **bronchioles contracts** and a large amount of **mucus** is produced.
3) This causes **constriction** of the airways, making it difficult for the sufferer to **breathe properly**. Air flow in and out of the lungs is **severely reduced**, so less oxygen enters the alveoli and moves into the blood.
4) Symptoms include **wheezing**, a **tight chest** and **shortness of breath**. During an attack the symptoms come on very suddenly. They can be relieved by **drugs** (often in **inhalers**) which cause the muscle in the bronchioles to **relax**, opening up the airways.

### Emphysema

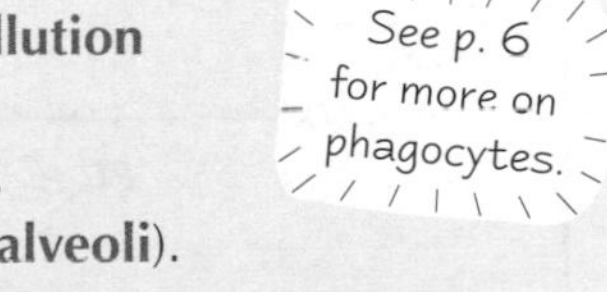


1) Emphysema is a lung disease caused by **smoking** or long-term exposure to **air pollution** — foreign particles in the smoke (or air) become **trapped** in the alveoli.
2) This causes **inflammation**, which attracts **phagocytes** to the area. The phagocytes produce an **enzyme** that breaks down **elastin** (a protein found in the **walls** of the **alveoli**).
3) Elastin is **elastic** — it helps the alveoli to **return** to their **normal shape** after inhaling and exhaling air.
4) Loss of elastin means the alveoli **can't recoil** to expel air as well (it remains **trapped** in the alveoli).
5) It also leads to **destruction** of the **alveoli walls**, which **reduces** the **surface area** of the alveoli so the rate of **gaseous exchange** decreases.
6) Symptoms of emphysema include **shortness of breath** and **wheezing**. People with emphysema have an **increased breathing rate** as they try to increase the amount of air (containing oxygen) reaching their lungs.

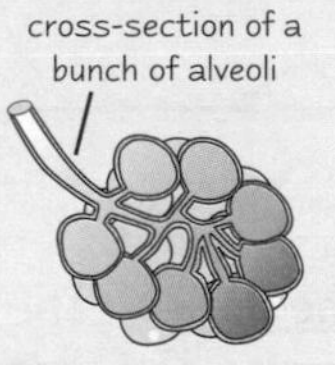


## *Practice Questions*

Q1 How is pulmonary ventilation calculated?

Q2 What happens to the alveoli of someone who suffers from emphysema?

**Exam Question**

Q1 Explain why a person suffering from fibrosis may feel tired and weak. [7 marks]

## *Asthma, emphysema and fibrosis — what a wheeze...*

*Tuberculosis is a pretty grim disease — all that coughing up blood and mucus. In total, around one third of the world's population (that's about two billion people) are infected. It's not just humans who are infected though — similar bacteria cause TB in loads of other animals, like cows, pigs, monkeys, goats, badgers, cats, dogs and children.*

# Interpreting Lung Disease Data

*It's very possible that you could be asked to interpret some data on lung disease in the exam. So being my usual nice self, I've given you some examples to show you how to do it. I know it looks a bit dull but believe me, it'll really help.*

## You Need to be Able to **Interpret Data** on **Risk Factors** and **Lung Disease**

1) All diseases have factors that will **increase** a person's **chance** of getting that disease. These are called **risk factors**. For example, it's widely known that if you **smoke** you're more likely to get **lung cancer** (smoking is a risk factor for lung cancer).
2) This is an example of a **correlation** — a link between two things (see page 90). However, a correlation doesn't always mean that one thing **causes** the other. Smokers have an **increased risk** of getting cancer but that doesn't necessarily mean smoking **causes** the disease — there are lots of other factors to take into consideration.
3) You need to be able to describe and analyse data given to you in your exams. Here are two examples of the kind of thing you might get:

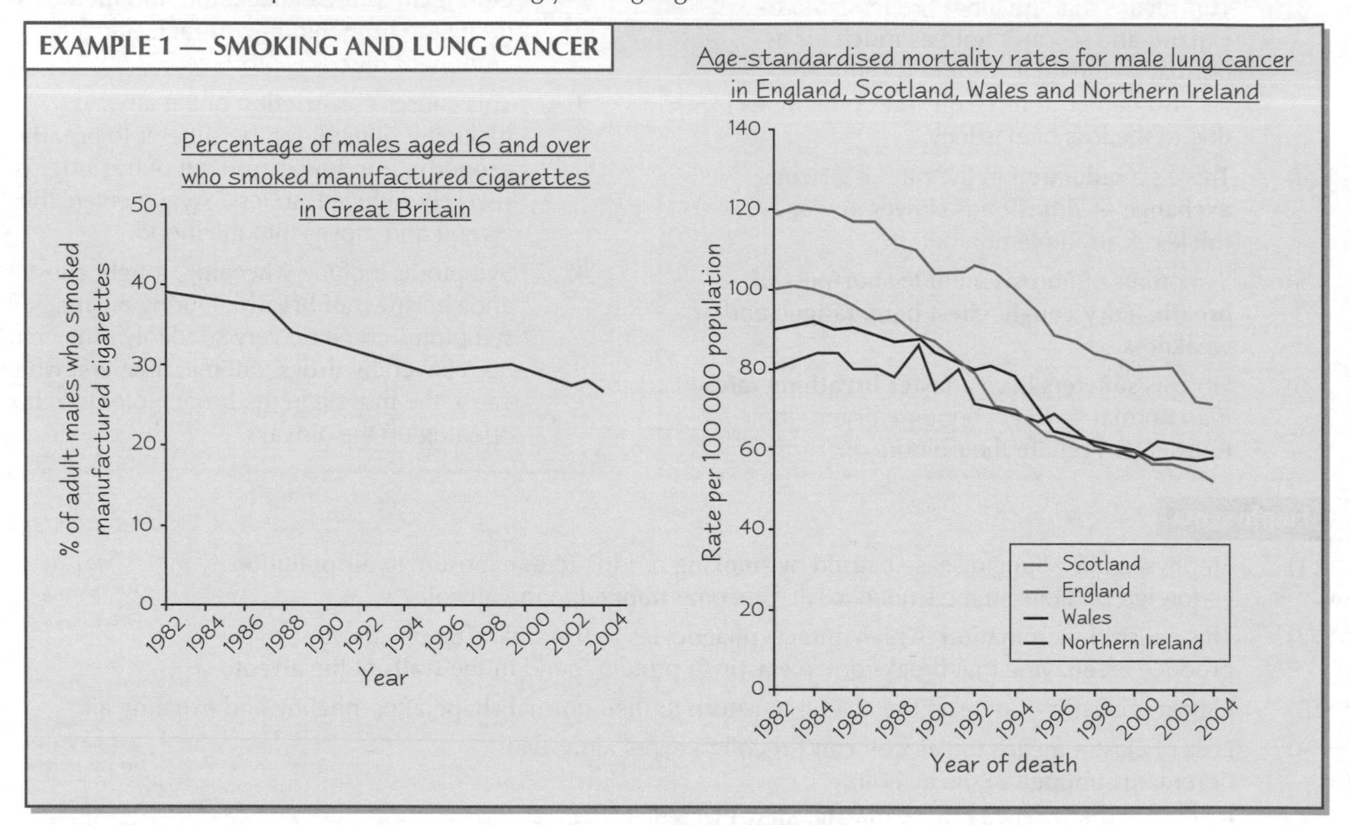


You might be asked to:

1) **Explain the data** — The graph on the left shows that the **number** of adult males in Great Britain (England, Wales and Scotland) who **smoke decreased** between 1982 and 2004. The graph on the right shows that the male lung cancer **mortality (death) rate decreased** between 1982 and 2004 for each of the countries shown. Easy enough so far.

See pages 90-92 for more on interpreting data.

2) **Draw conclusions** — You need to be careful what you say here. There's a **correlation** (link) between the **number** of males **who smoked** and the **mortality rate** for male lung cancer. But you **can't** say that one **caused** the other. There could be **other reasons** for the trend, e.g. deaths due to lung cancer may have decreased because less asbestos was being used in homes (not because fewer people were smoking).

**Other points to consider** — The graph on the right shows mortality (**death**) rates. The rate of **cases** of lung cancer **may have been increasing** but medical advances may mean more people were **surviving** (so only mortality was decreasing). Some information about the **people involved** in the studies would be helpful. For example, we don't know whether both studies used similar groups — e.g. similar diet, occupation, alcohol consumption etc. If they didn't then the results might not be reliable.

# Interpreting Lung Disease Data

## EXAMPLE 2 — AIR POLLUTION AND ASTHMA

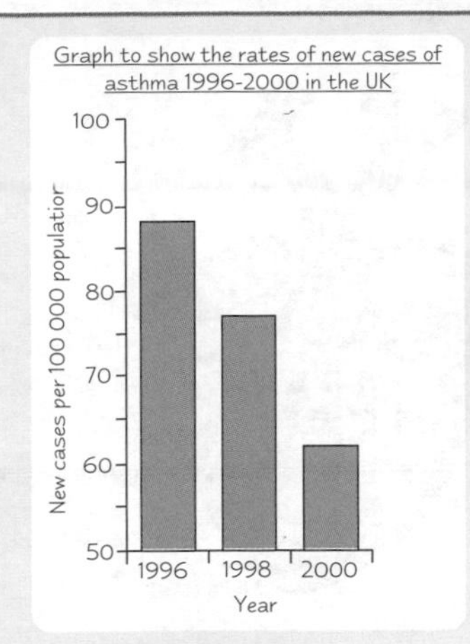


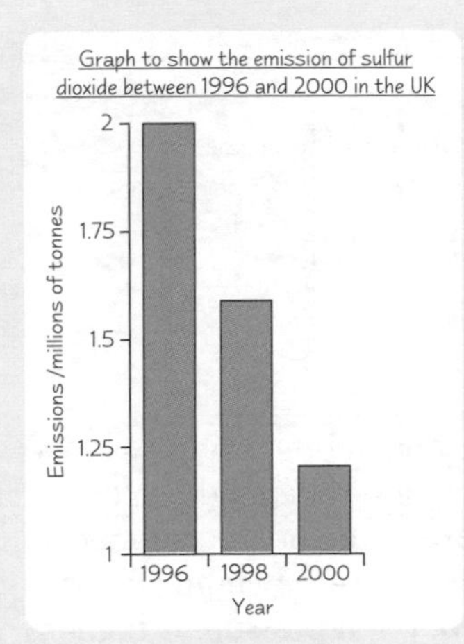


The **top graph** shows the number of **new cases of asthma** per 100 000 of the population diagnosed in the UK from 1996 to 2000. The **bottom graph** shows the **emissions** (in millions of tonnes) of **sulfur dioxide** (an **air pollutant**) from 1996 to 2000 in the UK.

**You might be asked to explain the data...**

1) The **top graph** shows that the number of **new cases of asthma** in the UK **fell** between 1996 and 2000, from 87 to 62 per 100 000 people.
2) The **bottom graph** shows that the **emissions of sulfur dioxide** in the UK **fell** between 1996 and 2000, from 2 to 1.2 million tonnes.

**... or draw conclusions**

1) Be careful what you say when drawing conclusions. Here there's a **link** between the **number** of new cases of **asthma** and **emissions of sulfur dioxide** in the **UK** — the rate of new cases of asthma has **fallen** as sulfur dioxide emissions have **fallen.** You **can't** say that one **causes** the other though because there could be **other reasons** for the trend, e.g. the number of new cases of asthma could be falling due to the **decrease** in the number of people **smoking.**
2) You can't say the **reduction** in asthma cases is **linked** to a **reduction in air pollution** (in general) either as **only** sulfur dioxide levels were studied.

**Other points to consider:**

1) The top graph shows **new cases** of asthma. The rate of new cases may be **decreasing** but existing cases may be becoming **more severe.**
2) The emissions were for the whole of the UK but air pollution **varies from area to area**, e.g. **cities** tend to be **more polluted.**
3) The asthma data doesn't take into account any **other factors** that may **increase** the risk of developing asthma, e.g. allergies, smoking, etc.

## Practice Exam Question

Q1 In early December 1952, a dense layer of cold air trapped pollutants close to ground level in London. The graph opposite shows daily deaths and levels of sulfur dioxide and smoke between 1st and 15th December.

a) Describe the changes in the daily death rate and the levels of pollutants over the days shown. [3 marks]

b) What conclusions can be drawn from this graph? [1 mark]

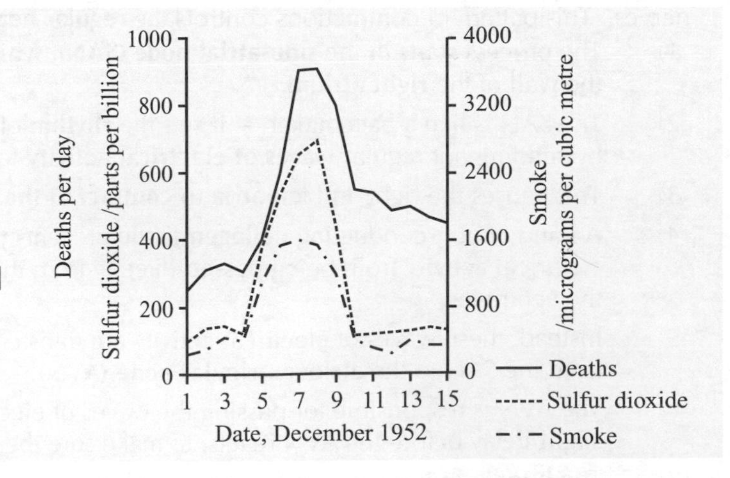


## Drawing conclusions — you'll need your wax crayons and some paper...

*These pages give examples to help you deal with what the examiners are sure to hurl at you — and boy, do they like throwing data around. There's some important advice here (even if I say so myself) — it's easy to leap to a conclusion that isn't really there — stick to your guns about the difference between correlation and cause and you'll blow 'em away.*

# The Heart

*The circulatory system is made up of the heart and blood vessels. Your heart is THE major player when it comes to circulating blood around your body — it's the 'pump' that gets oxygenated blood to your cells. So... unsurprisingly, you need to know how it works. You'll find that these pages definitely get to the heart of it... groan...*

## The **Heart** Consists of **Two Muscular Pumps**

The diagram on the right shows the **internal structure** of the heart. The **right side** pumps **deoxygenated blood** to the **lungs** and the **left side** pumps **oxygenated blood** to the **whole body**. Note — the **left and right sides** are **reversed** on the diagram, cos it's the left and right of the person that the heart belongs to.

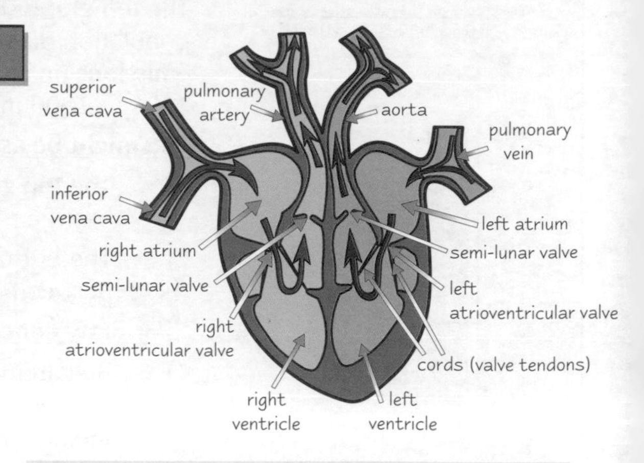


Each bit of the heart is adapted to do its job effectively.

1) The **left ventricle** of the heart has **thicker**, more muscular walls than the **right ventricle**, because it needs to contract powerfully to pump blood all the way round the body. The right side only needs to get blood to the lungs, which are nearby.
2) The **ventricles** have **thicker walls** than the **atria**, because they have to push blood out of the heart whereas the atria just need to push blood a short distance into the ventricles.
3) The **atrioventricular (AV) valves** link the atria to the ventricles and **stop blood flowing back** into the atria when the ventricles contract.
4) The **semi-lunar (SL) valves** link the ventricles to the pulmonary artery and aorta, and **stop blood flowing back** into the heart after the ventricles contract.
5) The **cords** attach the atrioventricular valves to the ventricles to stop them being forced up into the atria when the ventricles contract.

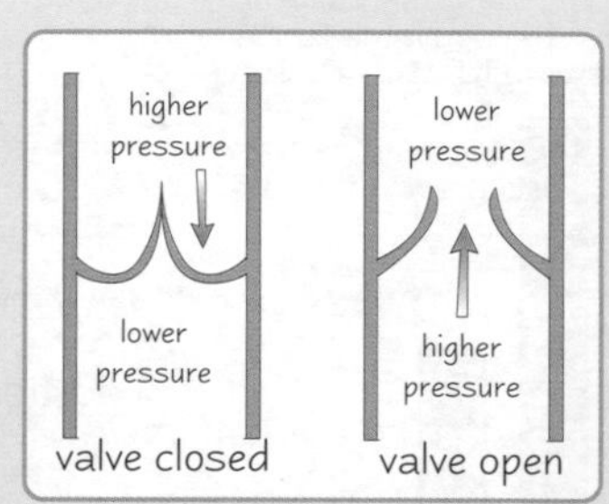


The **valves** only open one way — whether they're open or closed depends on the relative **pressure** of the heart chambers. If there's higher pressure **behind** a valve, it's forced **open**, but if pressure is higher **in front** of the valve it's forced **shut**.

## **Cardiac Muscle** Controls the **Regular Beating** of the Heart

Cardiac (heart) muscle is '**myogenic**' — this means that it can contract and relax without receiving signals from nerves. This pattern of contractions controls the **regular heartbeat**.

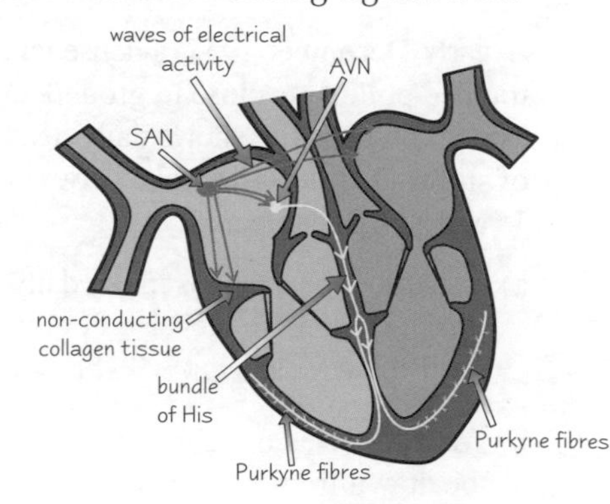


1) The process starts in the **sino-atrial node (SAN)**, which is in the wall of the **right atrium**.
2) The SAN is like a pacemaker — it sets the **rhythm** of the heartbeat by sending out regular **waves of electrical activity** to the atrial walls.
3) This causes the right and left **atria** to **contract at the same time**.
4) A band of non-conducting **collagen tissue** prevents the waves of electrical activity from being passed directly from the atria to the ventricles.
5) Instead, these waves of electrical activity are transferred from the SAN to the **atrioventricular node (AVN)**.
6) The AVN is responsible for passing the waves of electrical activity on to the bundle of His. But, there's a **slight delay** before the AVN reacts, to make sure the ventricles contract **after** the atria have emptied.
7) The **bundle of His** is a group of muscle fibres responsible for conducting the waves of electrical activity to the finer muscle fibres in the right and left ventricle walls, called the **Purkyne fibres**.
8) The Purkyne fibres carry the waves of electrical activity into the muscular walls of the right and left ventricles, causing them to **contract simultaneously**, from the bottom up.

# The Heart

## Learn the Equation for Cardiac Output

**Cardiac output** is the **volume** of blood pumped by the **heart per minute** (measured in $cm^3$ per minute).

It's calculated using this **formula**: **cardiac output = stroke volume × heart rate**

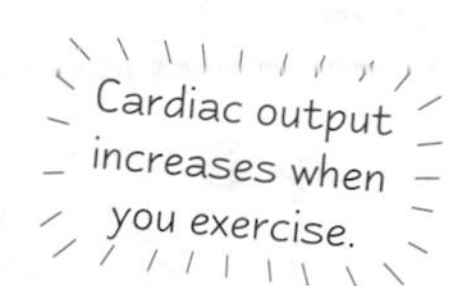


1) **Heart rate** — the **number** of **heartbeats** per minute. You can measure your heart rate by feeling your pulse, which is basically surges of blood forced through the arteries by the heart contracting.
2) **Stroke volume** — the **volume** of blood pumped during **each heartbeat**, measured in $cm^3$.

## The Cardiac Cycle Pumps Blood Round the Body

The cardiac cycle is an ongoing sequence of **contraction** and **relaxation** of the atria and ventricles that keeps blood **continuously** circulating round the body. The **volume** of the atria and ventricles **changes** as they contract and relax. **Pressure** changes also occur, due to the changes in chamber volume (e.g. decreasing the volume of a chamber by contraction will increase the pressure of a chamber). The cardiac cycle can be simplified into three stages:

Cardiac contraction is also called systole and relaxation is called diastole.

### 1) Ventricles relax, atria contract

The **ventricles are relaxed**. The **atria contract**, decreasing the volume of the chamber and **increasing** the **pressure** inside the chamber. This **pushes** the blood into the ventricles. There's a slight **increase** in **ventricular pressure** and **chamber volume** as the **ventricles receive the ejected blood** from the contracting atria.

### 2) Ventricles contract, atria relax

The **atria relax**. The **ventricles contract** (decreasing their volume), **increasing** their **pressure**. The pressure becomes **higher** in the ventricles than the atria, which forces the **AV valves shut** to prevent back-flow. The **pressure** in the **ventricles is also higher than in the aorta and pulmonary artery**, which forces **open** the **SL valves** and blood is forced out into these arteries.

### 3) Ventricles relax, atria relax

The **ventricles and the atria both relax**. The higher pressure in the pulmonary artery and aorta closes the SL valves to prevent back-flow into the ventricles. Blood returns to the heart and the **atria fill again** due to the higher pressure in the vena cava and pulmonary vein. In turn this starts to **increase** the **pressure** of the atria. As the ventricles continue to **relax**, their **pressure falls below the pressure of the atria** and so the **AV valves open**. This allows blood to flow **passively** (without being pushed by atrial contraction) into the ventricles from the atria. The atria contract, and the whole process begins again.

1 Atria contract
SL valves closed
vena cava
pulmonary vein
AV valves are open

2 Ventricles contract
SL valves forced open
blood leaves via pulmonary artery
blood leaves via aorta
AV valves forced closed

3 Atria and ventricles relax
SL valves forced closed
blood re-enters via vena cava
blood re-enters via pulmonary vein
AV valves forced open

There's a bit about interpreting cardiac cycle data on the next page. So turn over now — it's well exciting...

# The Heart

## You Might be Asked to *Interpret Data* on the *Cardiac Cycle*

You may well be asked to analyse or interpret **data** about the changes in **pressure** and **volume** during the cardiac cycle. Here are two examples of the kind of things you might get:

**Example 1**

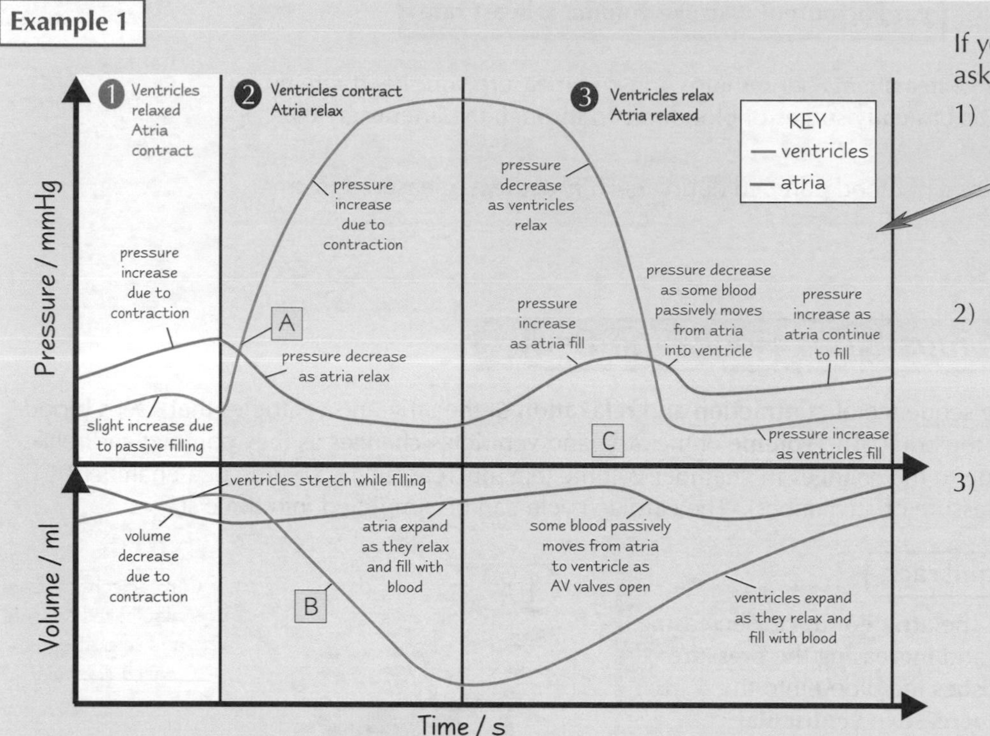


If you get a graph you could be asked **questions** like this:

1) **When** does blood start flowing into the **aorta**? At **point A**, the ventricles are **contracting** (and the AV valves are shut), forcing blood into the aorta.

2) Why is **ventricular volume decreasing** at **point B**? The ventricles are **contracting**, **reducing** the volume of the chamber.

3) Are the **semi-lunar valves** open or closed at **point C**? **Closed**. The ventricles are **relaxed** and **refilling**, so the pressure is **higher** in the **pulmonary artery** and **aorta**, forcing the SL valves **closed**.

The left ventricle has a thicker wall than the right ventricle and so it contracts more forcefully. This means the pressure is higher in the left ventricle (and in the aorta).

**Example 2**

You may have to describe the changes in pressure and volume shown by a **diagram**, like the one on the right. In this diagram the **AV valves** are **open**. So you know that the **pressure** in the **atria** is **higher** than in the **ventricles**. So you also know that the **atria are contracting** because that's what causes the **increase** in **pressure**.

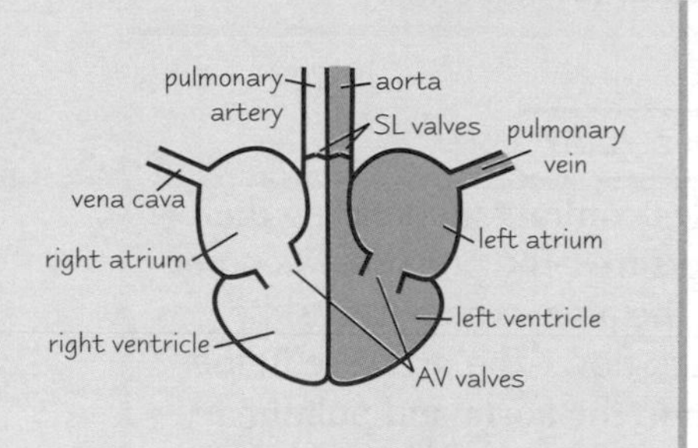


## Practice Questions

Q1 Which side of the heart carries oxygenated blood?

Q2 What does "myogenic" mean?

Q3 Describe the roles of the SAN and AVN.

These questions cover pages 40-42.

**Exam Questions**

Q1 Explain how valves in the heart stop blood going back the wrong way. [6 marks]

Q2 The table opposite shows the blood pressure in two heart chambers at different times during part of the cardiac cycle. Between what times:
a) are the AV valves shut? [1 mark]
b) do the ventricles start to relax? [1 mark]

| | Blood pressure / kPa | |
|---|---|---|
| Time / s | Left atrium | Left ventricle |
| 0.0 | 0.6 | 0.5 |
| 0.1 | 1.3 | 0.8 |
| 0.2 | 0.4 | 6.9 |
| 0.3 | 0.5 | 16.5 |
| 0.4 | 0.9 | 7.0 |

## My heart will go on... — well, I think I sound like Céline Dion anyway...

*Three whole pages to learn here, all full of really important stuff. If you understand all the pressure and volume changes then whether you get a diagram or a graph or something entirely different in the exam, you'll be able to interpret it, no probs.*

# Cardiovascular Disease

*No, your heart won't break if HE/SHE (delete as appropriate) doesn't return your call... but there are diseases associated with the heart and blood vessels that you have to learn...*

## *Most **Cardiovascular Disease** starts with **Atheroma** Formation*

1) The wall of an artery is made up of **several layers** (see p. 72).
2) The **endothelium** (inner lining) is usually smooth and unbroken.
3) If **damage** occurs to the endothelium (e.g. by high blood pressure), **white blood cells** (mostly macrophages) and **lipids** (fat) from the blood, clump together under the lining to form **fatty streaks**.
4) Over time, more white blood cells, lipids and **connective tissue** build up and harden to form a **fibrous plaque** called an **atheroma**.
5) This plaque **partially blocks** the lumen of the **artery** and **restricts blood flow**, which causes **blood pressure** to **increase**.

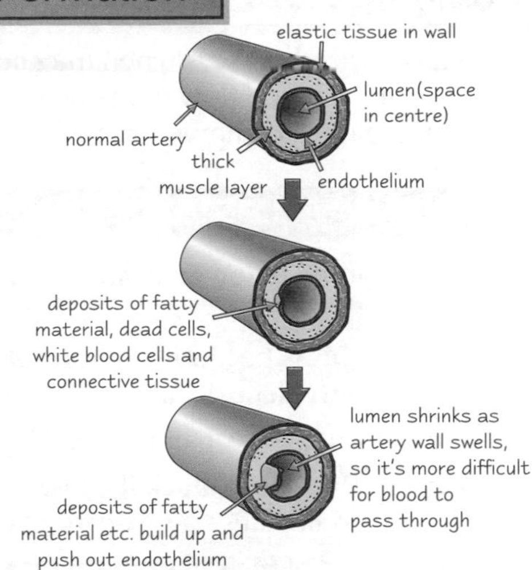


## *Atheromas Increase the Risk of Aneurysm and Thrombosis*

Two types of **disease** that affect the **arteries** are:

**Aneurysm** — a **balloon-like swelling** of the artery.

1) Atheroma plaques **damage** and **weaken arteries**. They also **narrow** arteries, **increasing blood pressure**.
2) When **blood** travels through a weakened artery at **high pressure**, it may **push** the **inner layers** of the artery **through the outer elastic layer** to form a **balloon-like swelling** — an **aneurysm**.
3) This aneurysm may **burst**, causing a **haemorrhage** (bleeding).

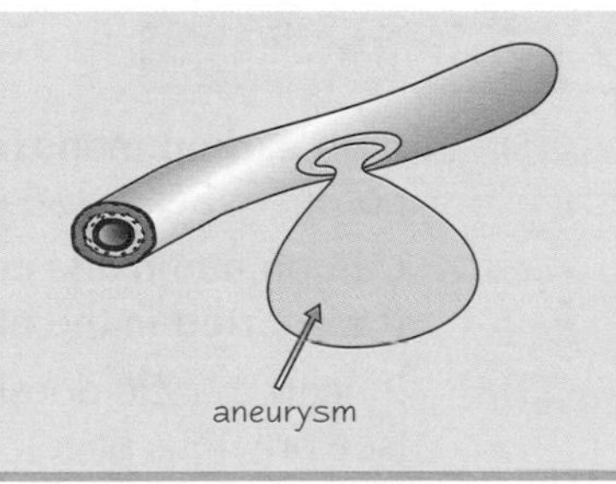


**Thrombosis** — formation of a **blood clot**.

1) An atheroma plaque can **rupture** (burst through) the **endothelium** (inner lining) of an artery.
2) This **damages** the artery wall and leaves a **rough** surface.
3) **Platelets** and **fibrin** (a protein) accumulate at the site of damage and form a **blood clot** (a thrombus).
4) This blood clot can cause a complete **blockage** of the artery, or it can become **dislodged** and block a blood vessel elsewhere in the body.
5) **Debris** from the rupture can cause another blood clot to form further down the artery.

## *Interrupted Blood Flow to the Heart can cause a Myocardial Infarction*

1) The **heart muscle** is supplied with **blood** by the **coronary arteries**.
2) This blood contains the **oxygen** needed by heart muscle cells to carry out **respiration**.
3) If a coronary artery becomes **completely blocked** (e.g. by a **blood clot**) an area of the heart muscle will be totally **cut off** from its blood supply, receiving **no oxygen**.
4) This causes a **myocardial infarction** — more commonly known as a **heart attack**.
5) A heart attack can cause **damage** and **death** of the **heart muscle**.
6) **Symptoms** include **pain** in the chest and upper body, **shortness of breath** and **sweating**.
7) If **large areas** of the heart are affected complete **heart failure** can occur, which is often **fatal**.

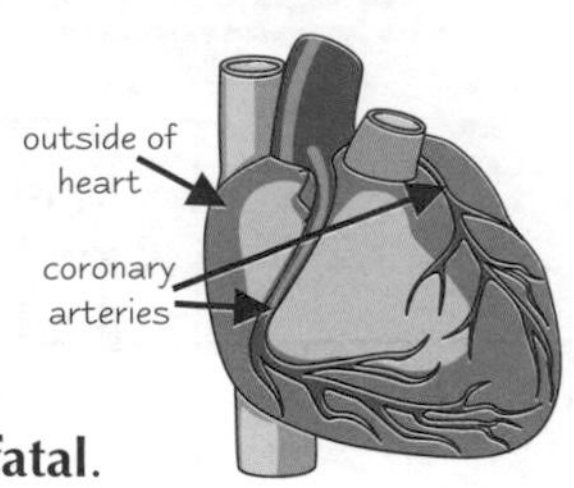

# Cardiovascular Disease

*There are treatments for cardiovascular disease out there, but it's best to try to avoid these diseases in the first place. Because lifestyle plays a large part, it's pretty easy to make preventive changes.*

## Some **Factors Increase** the **Risk** of **Coronary Heart Disease (CHD)**

**Coronary heart disease** is when the **coronary arteries** have lots of **atheromas** in them, which restricts blood flow to the heart. It's a type of **cardiovascular disease**. Some of the most common risk factors are:

### 1 High blood cholesterol and poor diet

1) If the **blood cholesterol level** is **high** (above 240 mg per 100 $cm^3$) then the risk of coronary heart disease is increased.
2) This is because **cholesterol** is one of the main constituents of the **fatty deposits** that form **atheromas** (see p. 43).
3) Atheromas can lead to **increased blood pressure** and **blood clots**.
4) This could **block** the flow of blood to **coronary arteries**, which could cause a **myocardial infarction** (see previous page for details).
5) A diet **high in saturated fat** is associated with high blood cholesterol levels.
6) A diet **high in salt** also **increases** the **risk** of cardiovascular disease because it increases the risk of **high blood pressure** (see below).

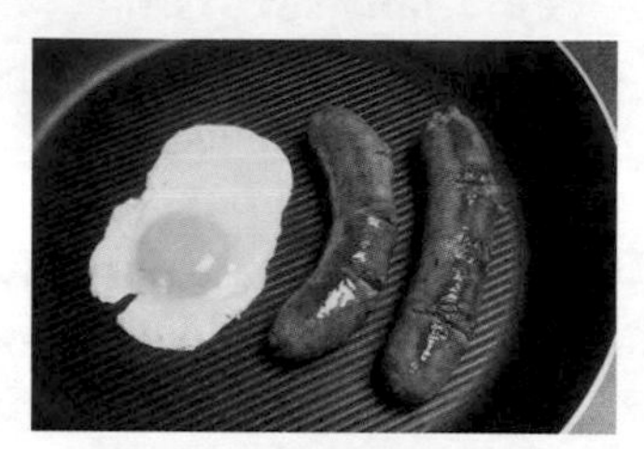

John decided to live on the edge and ordered a fry-up.

### 2 Cigarette smoking

1) Both **carbon monoxide** and **nicotine**, found in **cigarette smoke**, increase the risk of coronary heart disease.
2) Carbon monoxide combines with **haemoglobin** and **reduces** the amount of **oxygen transported** in the **blood**, and so reduces the amount of oxygen available to tissues.
3) If heart muscle doesn't receive enough oxygen it can lead to a **heart attack** (see previous page).
4) Smoking also **decreases** the **amount** of **antioxidants** in the blood — these are important for **protecting cells** from damage. Fewer antioxidants means **cell damage** in the **coronary artery walls** is more likely, and this can lead to **atheroma formation**.

### 3 High blood pressure

1) High blood pressure **increases** the **risk** of **damage** to the **artery walls**.
2) Damaged walls have an **increased risk** of **atheroma** formation, causing a further increase in blood pressure.
3) Atheromas can also cause **blood clots** to form (see p. 43).
4) A blood clot could **block flow** of **blood** to the heart muscle, possibly resulting in **myocardial infarction** (see p. 43).
5) So **anything** that **increases** blood pressure also increases the risk of **CHD**, e.g. being **overweight**, **not exercising** and excessive **alcohol** consumption.

Other factors include age (risk increases with age) and sex (men are more at risk than women).

Most of these factors are within our **control** — a person can **choose** to smoke, eat fatty foods, etc. However, some risk factors can't be controlled, such as having a **genetic predisposition** to coronary heart disease or having high blood pressure as a result of another **condition**, e.g. some forms of diabetes. Even so, the risk of developing CHD can be reduced by removing as many **risk factors** as you possibly can.

# Cardiovascular Disease

## You Might Have to *Interpret* Data on *Risk Factors* and *CHD*

Take a look at the following example of the sort of study you might see in your exam.

The graph shows the results of a study involving **34 439 male British doctors.** **Questionnaires** were used to find out the smoking habits of the doctors. The number of **deaths** among the participants from ischaemic heart disease (coronary heart disease) was counted, and **adjustments** were made to account for **differences in age.**

Here are some of the things you might be asked to do:

1) <u>Describe the data</u> — The **number** of deaths from ischaemic heart disease **increased** as the number of cigarettes smoked per day **increased. Fewer former smokers** and **non-smokers** died of ischaemic heart disease than smokers.
2) <u>Draw conclusions</u> — The **graph shows** a **positive correlation** between the number of cigarettes smoked per day by **male doctors** and the **mortality rate** from ischaemic heart disease.

Mortality rate from ischaemic heart disease per 1000 men /year
0
2
4
6
8
10
12
Never smoked
Former smoker
1-14
15-24
Over 24
Number of cigarettes smoked per day

3) <u>Explain the link</u> — You may get asked to **explain** the link between the risk factor under investigation (smoking) and CHD. E.g. **carbon monoxide** in **cigarette smoke** combines with **haemoglobin, reducing** the amount of **oxygen transported** in the **blood.** This **reduces** the amount of oxygen available to tissues, including **heart muscle**, which could lead to a **heart attack.** You would also talk about **antioxidants** (see previous page).

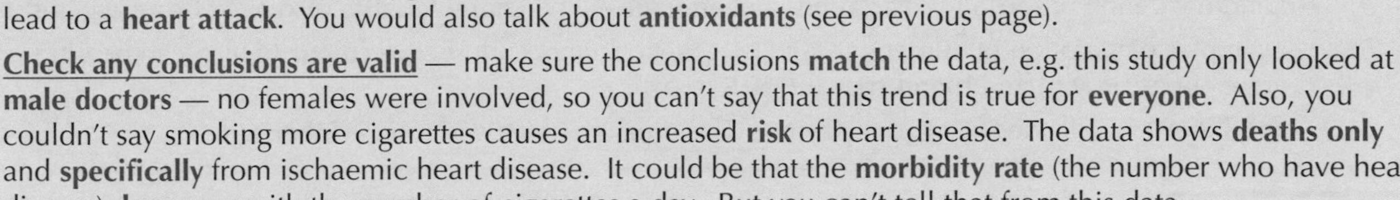

4) <u>Check any conclusions are valid</u> — make sure the conclusions **match** the data, e.g. this study only looked at **male doctors** — no females were involved, so you can't say that this trend is true for **everyone.** Also, you couldn't say smoking more cigarettes causes an increased **risk** of heart disease. The data shows **deaths only** and **specifically** from ischaemic heart disease. It could be that the **morbidity rate** (the number who have heart disease) **decreases** with the number of cigarettes a day. But you can't tell that from this data.

5) <u>Comment on the reliability of the results</u> — For example:
   - A **large sample size** was used — 34 439, which **increases** reliability.
   - People (even doctors) can tell **porkies** on questionnaires, **reducing** the **reliability** of results.

See pages 90-92 for more on interpreting data.

## Practice Questions

Q1 What is the name of the arteries that supply the heart muscle with oxygen?

Q2 What is the medical term for a heart attack?

Q3 What does CHD stand for?

Q4 Name three factors that can increase the chance of developing CHD.

**Exam Questions**

Q1 Describe how atheroma formation in the coronary arteries can lead to a myocardial infarction. [4 marks]

Q2 Describe how atheromas can increase the risk of a person suffering from an aneurysm. [3 marks]

## *Revision — increasing my risk of headache, stress, boredom...*

*I know there's a lot to take in on these pages... but make sure you understand the link between atheromas, thrombosis and heart attacks — basically an atheroma forms, which can cause thrombosis, which can lead to a heart attack. Anything that increases the chance of an atheroma forming (high blood pressure, smoking, fatty diet) is bad news for your heart...*

# Causes of Variation

*Every organism — man, woman, gibbon or platypus — differs in some way from every other. The differences between organisms is called variation. It's all down to genes and the environment.*

## ***Variation*** Exists **Between All Individuals**

**Variation** is the **differences** that exist between **individuals**. There are two types:

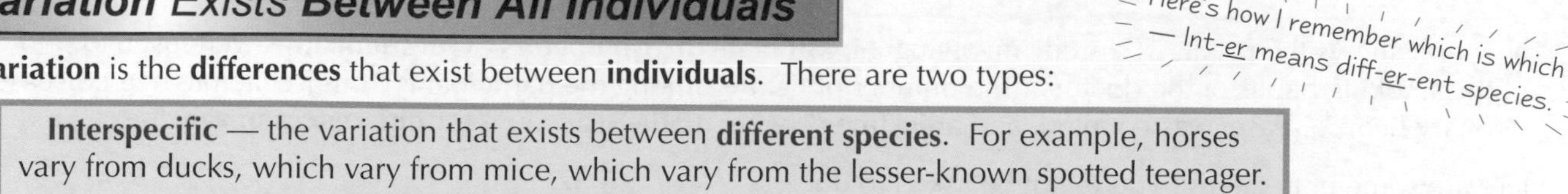


**Interspecific** — the variation that exists between **different species**. For example, horses vary from ducks, which vary from mice, which vary from the lesser-known spotted teenager.

**Intraspecific** — the differences that occur **within a species**. For example, the number of 'eyes' on peacocks' feathers, or the length of giraffes' necks.

## ***Intraspecific Variation*** is Caused by ***Genetic*** and ***Environmental Factors***

Although individuals of the same species may **appear similar**, no two individuals are **exactly alike**. **Variation** can be caused by both **genetic** and **environmental** factors.

**Genetic**

1) All the members of a species have the **same genes** — that's what makes them the **same species**.
2) But **individuals** within a species can have **different versions** of those genes — called **alleles** (see p. 53).
3) The alleles an organism has make up its **genotype**.
4) Different genotypes result in **variation** in **phenotype** — the **characteristics** displayed by an organism.
5) Examples of variation in humans caused by genetic factors include **eye colour** (which can be blue, green, grey, brown) and **blood type** (O, A, B or AB).
6) You **inherit** your genes from your parents. This means genetic variation is **inherited**.

**Environmental**

The **appearance** (**phenotype**) of an individual is also affected by the **environment**, for example:

1) **Plant growth** is affected by the amount of **minerals**, such as **nitrate** and **phosphate**, available in the soil.
2) **Fur colour** of the **Himalayan rabbit** is affected by **temperature**. Most of the rabbits' fur is **white** except the ears, feet and tail, which are **black**. The black colour only develops in temperatures **below 25 °C**. If a patch of their white fur is **shaved** and a cold pad applied to the shaved area, the hair that grows back will be **black**.
3) **Identical twins** are **genetically** identical — they have the same alleles, so any differences between them will be due to the **environment**. For example, they may have had different **illnesses** that affected their **development** or, if they grew up in **different areas**, they may have different **accents**.

Thumper and Biggles went to any length to not touch the cold floor.

## ***Variation*** is Often a **Combination** of ***Genetic*** and ***Environmental Factors***

An individual may have the **genetic information** for a **particular characteristic**, but **environmental factors** may affect the **expression** of this characteristic, for example:

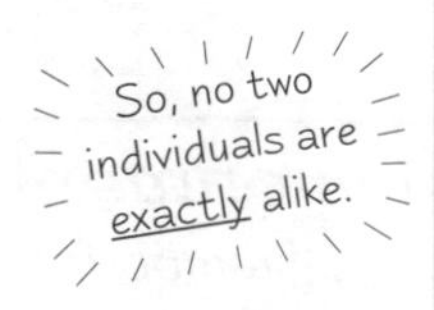


1) A person might have the **genes** to potentially grow to be **six foot tall**. Whether or not they grow to this height will **depend** on **environmental factors** such as their **diet** and **health**.
2) The amount of **melanin** (pigment) people have in their **skin** is partially controlled by their **genes**, but skin colour is influenced by the **amount of sunlight** a person is exposed to.

# Causes of Variation

## Be Careful When Drawing Conclusions About the Cause of Variation

In any **group of individuals** there's a lot of **variation** — think how different all your friends are. It's not always **clear** whether the variation is caused by **genes**, the **environment** or **both**. Scientists draw conclusions based on the information they have until **new evidence** comes along that **challenges it** — have a look at these two examples:

**Example 1 — Overeating**

1) **Overeating** was thought to be caused only by environmental factors, like an **increased availability of food** in developed countries.
2) It was later discovered that food consumption **increases** brain **dopamine** levels in animals.
3) Once enough dopamine was released, people would **stop** eating.
4) Researchers discovered that people with one particular **allele** had **30% fewer** dopamine receptors.
5) They found that people with this particular allele were **more likely** to overeat — they wouldn't stop eating when dopamine levels increased.
6) Based on this evidence, scientists now think that overeating has **both genetic** and **environmental** causes.

**Example 2 — Antioxidants**

1) Many foods in our diet contain **antioxidants** — compounds that are thought to play a role in **preventing chronic diseases**.
2) Foods such as **berries** contain **high levels** of antioxidants.
3) Scientists thought that the berries produced by different **species** of plant contained **different levels** of antioxidants because of **genetic factors**.
4) But experiments that were carried out to see if **environmental** conditions affected antioxidant levels found that environmental conditions caused a great deal of **variation**.
5) Scientists now believe that antioxidant levels in berries are due to **both genetic** and **environmental** factors.

## Practice Questions

Q1 What is intraspecific variation?

Q2 What is interspecific variation?

Q3 Which two kinds of factors can cause variation?

Q4 Give one example of a characteristic that varies due to both genetic and environmental factors.

**Exam Question**

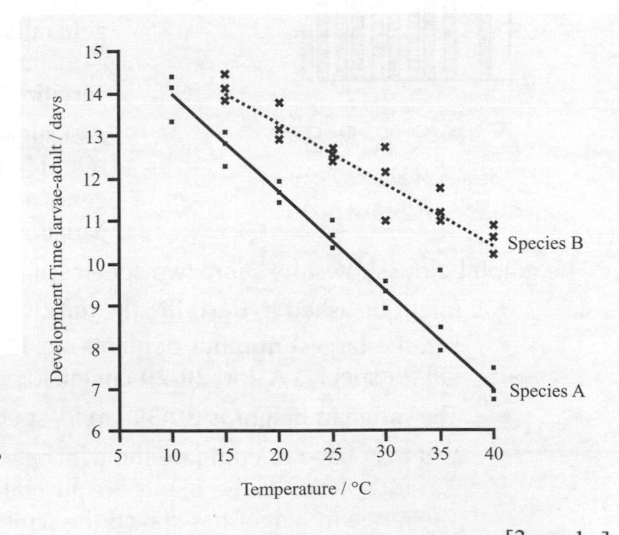


Q1 The graph shows the results of an investigation into the effects of temperature on the length of time it took for ladybird larvae to emerge as adults. Two species of ladybird were investigated, species A and species B.

a) Describe the results of the study. [3 marks]

b) Explain what causes the variation between the species and within each species. [4 marks]

## Environmental Factor — the search is on for the most talented environment...

*Err... Inter... Intra... I don't know who thought that using two words almost exactly the same was a good idea, but it wasn't very helpful. Good thing I've made these pages then I reckon. Copy out the definitions a few times and make sure you understand the examples — they'll use the same ideas but different case studies in the exam.*

# Investigating Variation

*It's a lot of work studying variation in an entire population (imagine studying all the ants in one nest) — so instead you can take a random sample and use this to give you a good idea of what's going on in the entire population.*

## To **Study** Variation You Have to **Sample** a **Population**

When studying variation you usually only look at a **sample** of the population, **not** the **whole thing**. For most species it would be too **time-consuming** or **impossible** to catch all the individuals in the group. So samples are used as **models** for the **whole population**.

## The **Sample** has to be **Random**

Because sample data will be used to **draw conclusions** about the **whole population**, it's important that it **accurately represents** the whole population and that any patterns observed are tested to make sure they're not due to chance.

1) To make sure the sample isn't **biased**, it should be **random**. For example, if you were looking at plant species in a field you could pick random sample sites by dividing the field into a **grid** and using a **random number generator** to select coordinates.
2) To ensure any variation observed in the sample isn't just due to **chance**, it's important to analyse the results **statistically**. This allows you to be more **confident** that the results are true and therefore will reflect what's going on in the **whole population**.

## You Need to be Able to **Analyse** and **Interpret Data** Relating to **Variation**

You might be asked to **analyse** and **interpret** data relating to **interspecific and intraspecific variation** in your exam. So here's a big **example** to give you an idea of what you might get:

The graph below shows the growth of two **different** species of plant in the **same environment**. You might be asked to:

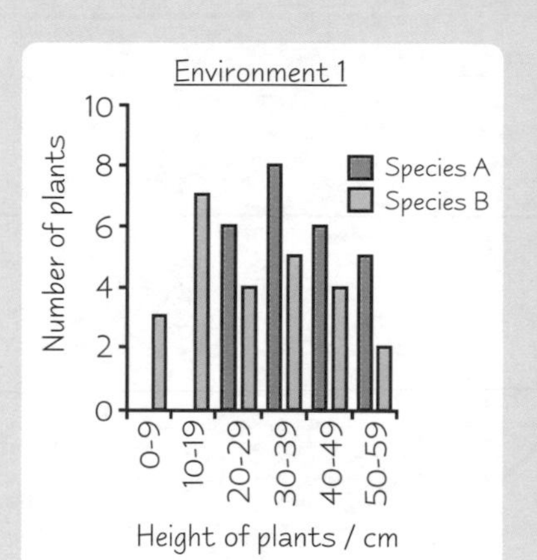


1) **Describe the data**...
   - The largest number of plants are **30-39 cm** tall for species **A** and **10-19 cm** tall for species **B**.
   - Species **A** plants **range in height** from **20-59 cm** but the range is **larger** for **species B** (**0-59 cm**).
2) ...or **draw conclusions**
   There is **interspecific variation** in plant height — Species **A** plants are **generally taller** than Species B. **Both species** show **intraspecific variation** — plant height varies for both species. There is **more intraspecific variation** in species **B** — the range of heights is bigger.
3) ...or **suggest a reason** for the differences
   A and B are **separate species**, grown in the same area. This means their **genes are different** but their **environment is the same**. So any **interspecific** variation in height is down to **genetic factors**, **not** the environment.

The graph below shows the **same** two species of plant but grown in a **different environment** than in the first graph.

1) You might be asked to **describe the data**...
   - E.g. the **largest number** of plants are **40-49 cm** tall for species **A** and **20-29 cm** tall for species **B**.
   - The **range in height** is **20-59 cm** for species **A** and **10-59 cm** for species **B**.
   - You may have to **compare** the data between the two graphs. E.g. for species **A**, the plants are generally **taller** in **environment 2**. The range in height has **stayed the same**. For species **B**, the plants are also generally **taller** in **environment 2**, but the range in height is **smaller**.
2) ...or **draw conclusions and suggest a reason** for the differences
   - Both species are **generally taller** in environment **2** than in environment 1. So, variation in height is affected by **environmental factors**.
   - Species A shows **similar** height variation in **both environments**. The variation in species B **differs** between the two environments. This suggests that **environmental factors influence** height **more** in species **B**.

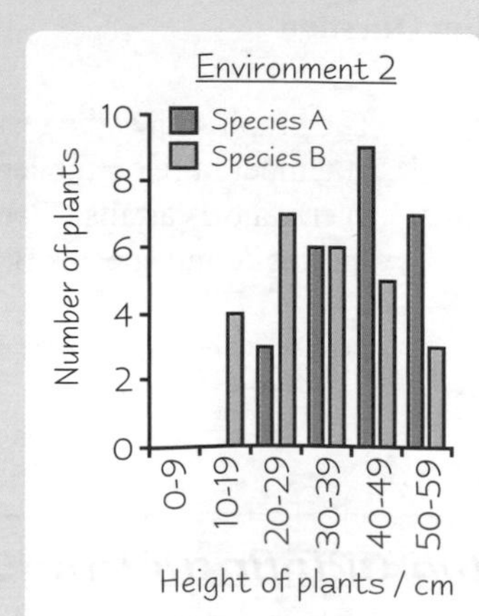

# Investigating Variation

## **Standard Deviation** Tells You About the Variation **Within a Sample**

1) The **mean value** tells you the **average** of the values collected in a **sample**.
2) It can be used to tell if there **is variation between samples**, e.g. the mean number of apples produced by species A = 26 and B = 32. So the **number** of apples produced by different tree species **does vary**.
3) Most samples give you a **bell-shaped graph** — this is called a **normal distribution**.
4) The **standard deviation** tells you **how much** the values in a **single sample vary**. It's a **measure** of the **spread of values about the mean**.
5) For example:
   - Species A: mean = 26, standard deviation = 3 — most of the trees in the sample produced between 23 and 29 apples (26 ± 3).
   - Species B: mean = 32, standard deviation = 9 — most of the trees in the sample produced between 23 and 41 apples (32 ± 9).
   - So species B generally produces **more apples** but shows a **greater variation** in the number produced, compared to species A.

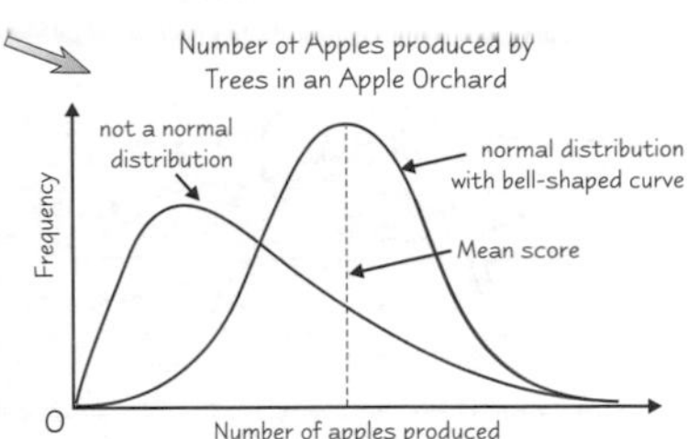


6) A **large standard deviation** means the values in the sample **vary a lot**. A **small standard deviation** tells you most of the sample data is around the mean value, so **varies little**:

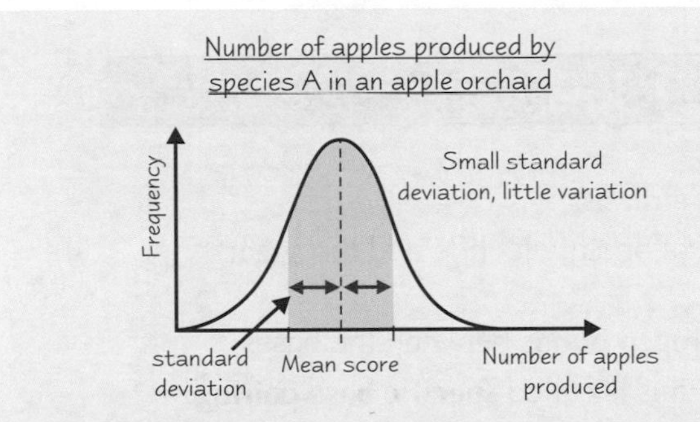


When all the values are **similar**, so vary little, the graph is **steep** and the standard deviation is **small**.

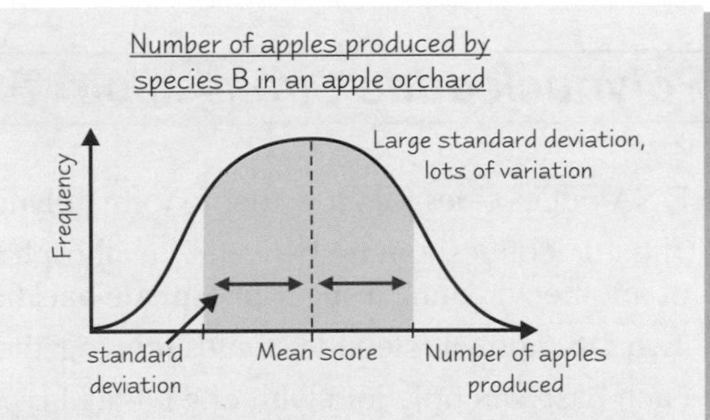


When all the values vary a lot, the graph is **fatter** and the standard deviation is **large**.

## Practice Questions

Q1 Why do scientists look at a sample of a population, rather than the whole population?

Q2 Why does a population sample have to be chosen at random?

Q3 What does the standard deviation of a data set tell us?

**Exam Question**

Q1 A study was conducted into how smoking during pregnancy affects the birth mass of newborn babies, depending on the genotype of the mother. The results showed that women who smoked during the entire pregnancy had babies with a mean reduction in birth mass of 377 grams. But the reduction was as much as 1285 grams among women with certain genotypes.

a) Data on variation in child birth mass was also collected from a group of non-smokers. Suggest why this data was collected. [1 mark]

b) What can be concluded about the influence of genetic factors and environmental factors on birth mass? Give evidence from the study to support your answer. [4 marks]

c) Give two other factors that should be controlled in this experiment. [2 marks]

## Investigating standards — say that to your teachers to scare them...

*Bet you thought you'd finished with maths — 'fraid not. Thankfully you don't need to know how to work out standard deviation though. But you do need to know how to go about interpreting data — which is a bit more interesting than maths, and I can exclusively reveal it's what all the examiners dream about at night.*

# DNA

*These pages are about the wonderful world of genetics. But you won't be able to make head or tail of it without learning all about DNA (**d**eoxyribo**n**ucleic **a**cid), so here goes...*

## DNA is Made of Nucleotides that Contain a Sugar, a Phosphate and a Base

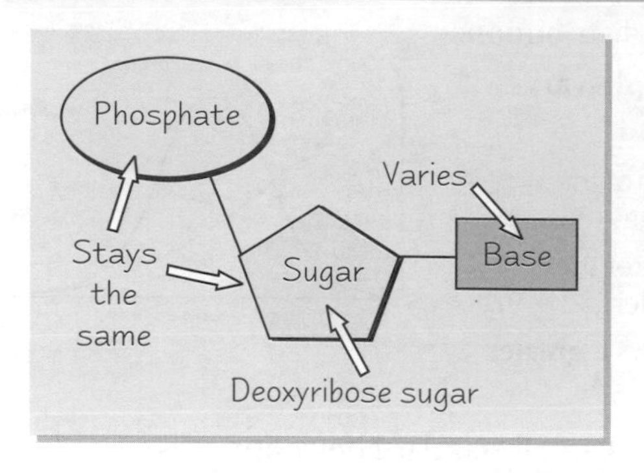


1) DNA is a polynucleotide — it's made up of lots of **nucleotides** joined together.
2) Each nucleotide is made from a **pentose sugar** (with 5 carbon atoms), a **phosphate** group and a **nitrogenous base**.
3) The **sugar** in DNA nucleotides is a **deoxyribose** sugar.
4) Each nucleotide has the **same sugar and phosphate**. The **base** on each nucleotide can **vary** though.
5) There are **four** possible bases — adenine (**A**), thymine (**T**), cytosine (**C**) and guanine (**G**).

## Two Polynucleotide Strands Join Together to Form a Double-Helix

1) DNA nucleotides join together to form **polynucleotide strands**.
2) The nucleotides join up between the **phosphate** group of one nucleotide and the **sugar** of another, creating a **sugar-phosphate backbone**.
3) **Two** DNA polynucleotide strands join together by **hydrogen bonds** between the bases.
4) Each base can only join with one particular partner — this is called **specific base pairing**.
5) **Adenine** always pairs with **thymine** (**A - T**) and **guanine** always pairs with **cytosine** (**G - C**).
6) The two strands **wind up** to form the **DNA double-helix**.

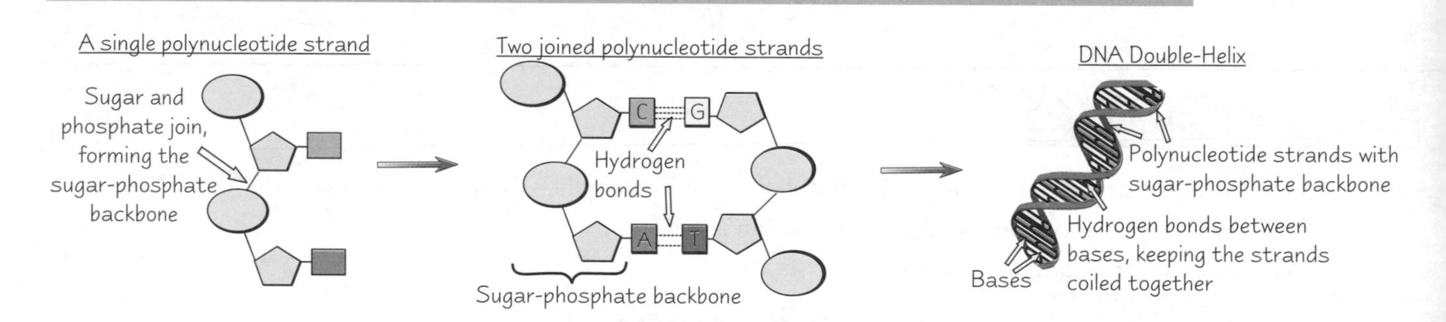


## DNA's Structure Makes It Good at Its Job

1) DNA contains your **genetic information** — that's **all the instructions** needed to **grow and develop** from a fertilised egg to a fully grown adult.
2) The DNA molecules are very **long** and are **coiled** up very tightly, so a lot of genetic information can fit into a **small space** in the cell nucleus.
3) DNA molecules have a **paired structure**, which makes it much easier to **copy itself**. This is called **self-replication** (see p. 62). It's important for cell division (see p. 64) and for passing genetic information from **generation to generation** (see p. 54).
4) The double-helix structure means DNA is **very stable** in the cell.

Geoff's helix wasn't as tightly coiled as DNA but it was a lot more fun.

# DNA

## DNA is Stored Differently in Different Organisms

Although the **structure** of DNA is the same in all organisms, **eukaryotic** and **prokaryotic** cells store DNA in slightly different ways. (For a recap on the differences between prokaryotic and eukaryotic cells see p. 32.)

### Eukaryotic DNA is Linear and Associated with Proteins

1) Eukaryotic cells contain **linear** DNA molecules that exist as **chromosomes** — thread-like structures, each made up of **one long molecule** of DNA.
2) The DNA molecule is **really long** so it has to be **wound up** so it can **fit** into the nucleus.
3) The DNA molecule is wound around **proteins** (called **histones**).
4) Histone proteins also help to **support** the DNA.
5) The DNA (and protein) is then coiled up **very tightly** to make a **compact chromosome**.

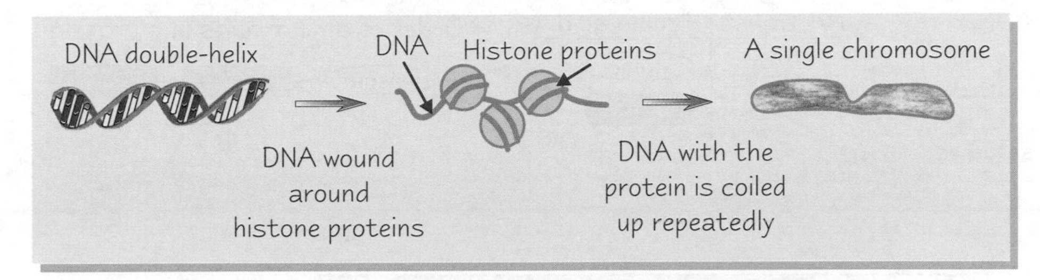


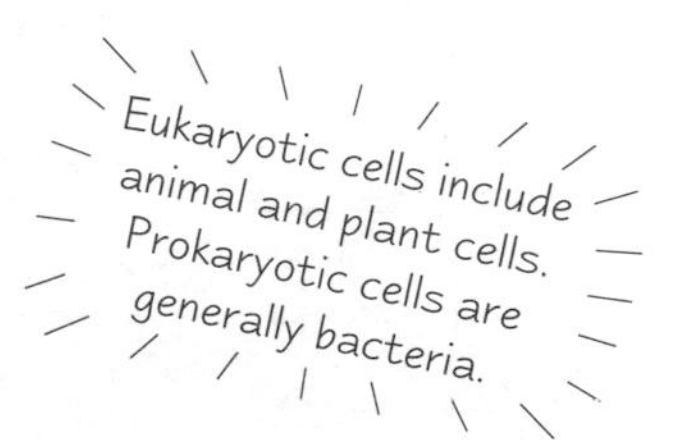


### DNA Molecules are Shorter and Circular in Prokaryotes

1) Prokaryotes also carry DNA as **chromosomes** — but the DNA molecules are **shorter** and **circular**.
2) The DNA **isn't** wound around proteins — it condenses to fit in the cell by **supercoiling**.

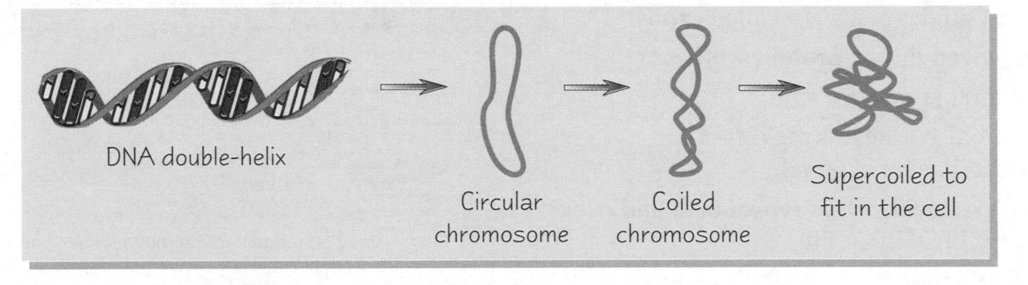

If one more person confused Clifford with supercoiled DNA, he'd have 'em.

## Practice Questions

Q1 What are the three main components of nucleotides?

Q2 Which bases join together in a DNA molecule?

Q3 What type of bonds join the bases together?

Q4 Why is DNA so tightly coiled?

**Exam Questions**

Q1 Describe, using diagrams where appropriate, how nucleotides are joined together in DNA and how two single polynucleotide strands of DNA are joined. [4 marks]

Q2 Describe how DNA is stored in prokaryotic and eukaryotic cells. [5 marks]

## Give me a D, give me an N, give me an A! What do you get? — Very confused...

*You need to learn the structure of DNA — the sugar-phosphate backbone, the hydrogen bonds, and don't forget the base pairing. Then there's the differences in the way that DNA is stored in eukaryotic and prokaryotic cells. Pheww.*

# Genes

*Now you've got to grips with the structure of DNA, you can learn how DNA is used to carry information. It's all in the sequence of bases you see...*

## DNA Contains Genes Which are Instructions for Proteins

1) Genes are **sections of DNA**. They're found on **chromosomes**.
2) Genes **code** for **proteins** (polypeptides) — they contain the **instructions** to make them.
3) Proteins are made from **amino acids**.
4) Different proteins have a **different number** and **order** of amino acids.
5) It's the **order** of **nucleotide bases** in a gene that determines the **order of amino acids** in a particular **protein**.
6) Each amino acid is coded for by a sequence of **three bases** (called a **triplet**) in a gene.
7) Different sequences of **bases** code for different **amino acids**. For example:

Polypeptide is just another word for a protein.

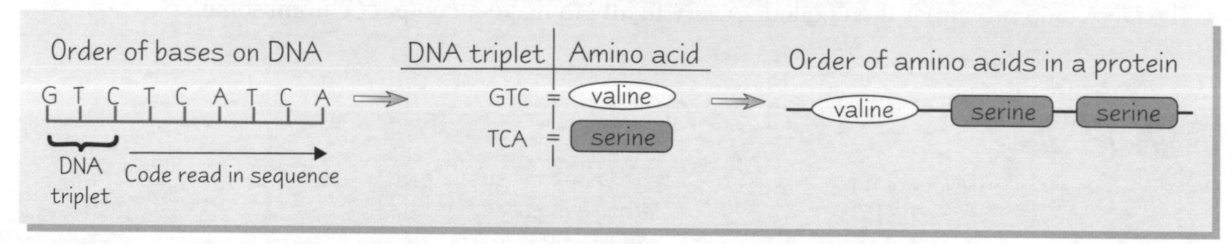


## Not All the DNA in Eukaryotic Cells Codes for Proteins

1) Genes in eukaryotic DNA contain sections that **don't code** for amino acids.
2) These sections of DNA are called **introns** (all the bits that do code for amino acids are called **exons**).
3) Introns are **removed** during **protein synthesis**. Their purpose isn't known for sure.
4) Eukaryotic DNA also contains regions of **multiple repeats** outside of genes.
5) These are DNA sequences that **repeat** over and over. For example: CCTTCCTTCCTT.
6) These areas **don't code** for amino acids either.

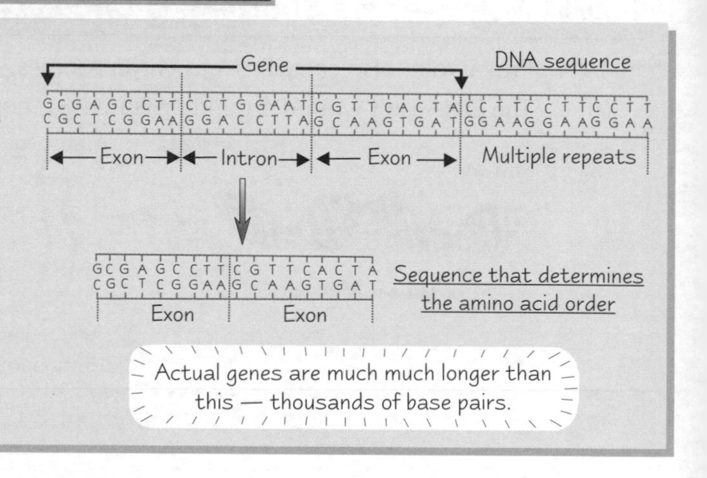


Actual genes are much much longer than this — thousands of base pairs.

## The Nature and Development of Organisms is Determined by Genes

1) **Enzymes** speed up most of our **metabolic pathways** — the chemical reactions that occur in the body. These pathways determine how we **grow and develop**.
2) Because enzymes control the metabolic pathways, they **contribute** to our **development**, and ultimately what we look like (our **phenotype**).
3) All enzymes are **proteins**, which are built using the **instructions** contained within genes. The **order of bases** in the gene decides the order of **amino acids** in the protein and so what type of protein (or enzyme) is made.
4) So, our genes help to **determine** our **nature**, **development** and **phenotype** because they contain the information to **produce** all our proteins and enzymes.

Ken's bad fashion sense was literally down to his genes. (A genes/jeans joke — classic CGP.)

This flowchart shows how **DNA** determines our **nature** and **development**:

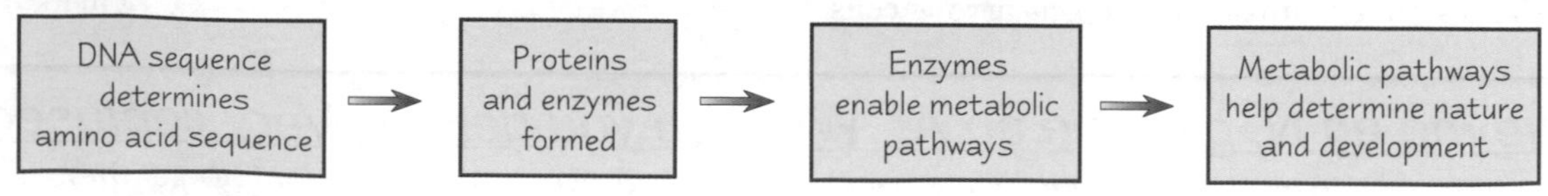

# Genes

## Genes can Exist in Different Forms Called Alleles

1) A gene can exist in more than one form. These forms are called **alleles**.
2) The order of bases in each allele is slightly different, so they code for **slightly different versions** of the **same characteristic**. For example, the gene that codes for **blood type** exists as one of three alleles — one codes for type O, another for type A and the other for type B.

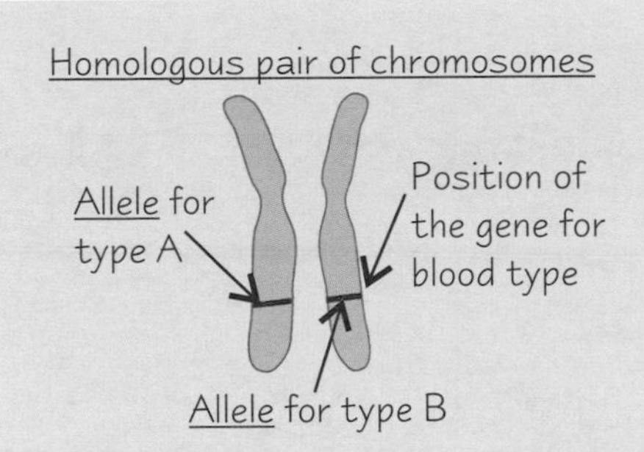


Our DNA is stored as **chromosomes** in the nucleus of cells. Humans have **23 pairs** of chromosomes, 46 in total — two number 1s, two number 2s, two number 3s etc. Pairs of matching chromosomes (e.g. the 1s) are called **homologous pairs**. In a homologous pair both chromosomes are the same size and have the **same genes**, although they could have **different alleles**. Alleles coding for the same characteristic will be found at the **same position** (**locus**) on each chromosome in a homologous pair.

## Gene Mutations can Result in Non-functioning Proteins

1) **Mutations** are **changes** in the **base sequence** of an organism's **DNA**.
2) So, mutations can produce **new alleles** of genes.
3) A gene codes for a particular protein, so if the sequence of bases in a gene changes, a **non-functional** or **different protein** could be produced.
4) All **enzymes** are **proteins**. If there's a mutation in a gene that codes for an enzyme, then that enzyme may not **fold up** properly. This may produce an **active site** that's the wrong shape and so a **non-functional enzyme**.

See p. 18 for a recap of enzymes and active sites.

**Example**

Gene X codes for an enzyme that catalyses the conversion of A to B. A mutation in the gene may result in the formation of a non-functional enzyme. This means that the reaction **can't happen**.

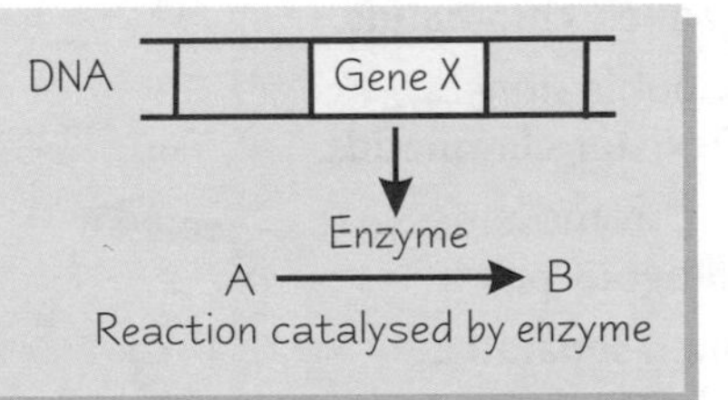


## Practice Questions

Q1 What is a DNA triplet?

Q2 What is an intron?

Q3 What is an allele?

Q4 How can a mutation result in a non-functional enzyme?

| Amino acid | DNA triplet |
|---|---|
| Glycine | GGC |
| Glutamic acid | GAG |
| Proline | CCG |
| Tryptophan | TGG |

**Exam Questions**

Q1 a) Write a definition of a gene. [2 marks]

b) Use the table above to write the protein sequence coded for by the DNA sequence TGGCCGCCGGAG. [1 mark]

Q2 Describe how the DNA of an organism helps to determine its nature and development. [4 marks]

## Exons stay in, introns go out, in out, in out, and shake it all about...

*Quite a few terms to learn here I'm afraid. Some are a bit confusing too. Just try to remember which way round they go. Introns are the non-coding regions but exons are extremely important — they actually code for the protein (polypeptide).*

# Meiosis and Genetic Variation

*Humans are all similar because we have the same genes. But we show genetic variation because we inherit different combinations of alleles from our parents. This is what leads to the differences you can see between you and your siblings, and your friends, and your postman... (unless your mum's not told you something...)*

## DNA from One Generation is Passed to the Next by Gametes

1) **Gametes** are the **sperm** cells in males and **egg** cells in females. They join together at **fertilisation** to form a **zygote**, which divides and develops into a **new organism**.
2) Normal **body cells** have the **diploid number** (**2n**) of chromosomes — meaning each cell contains **two** of each chromosome, one from the mum and one from the dad.
3) **Gametes** have a **haploid** (**n**) number of chromosomes — there's only one copy of each chromosome.
4) At **fertilisation**, a **haploid sperm** fuses with a **haploid egg**, making a cell with the normal diploid number of chromosomes. Half these chromosomes are from the father (the sperm) and half are from the mother (the egg).

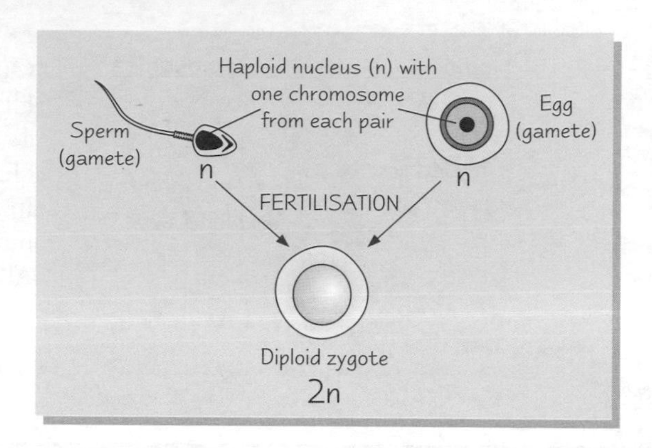


## Gametes are Formed by Meiosis

**Meiosis** is a type of cell division. Cells that divide by meiosis are **diploid** to start with, but the cells that are formed from meiosis are **haploid** — the chromosome number **halves**. Without meiosis, you'd get **double** the number of chromosomes when the gametes fused. Not good.

1) The DNA unravels and **replicates** so there are **two** copies of **each** chromosome, called **chromatids**.
2) The DNA condenses to form double-armed chromosomes, made from **two sister chromatids**.
3) **Meiosis I** (first division) — the chromosomes arrange themselves into **homologous pairs**.
4) These homologous **pairs** are then **separated**, **halving** the chromosome number.
5) **Meiosis II** (second division) — the pairs of sister **chromatids** that make up each chromosome are **separated**.
6) **Four haploid cells** (gametes) that are **genetically different** from each other are produced.

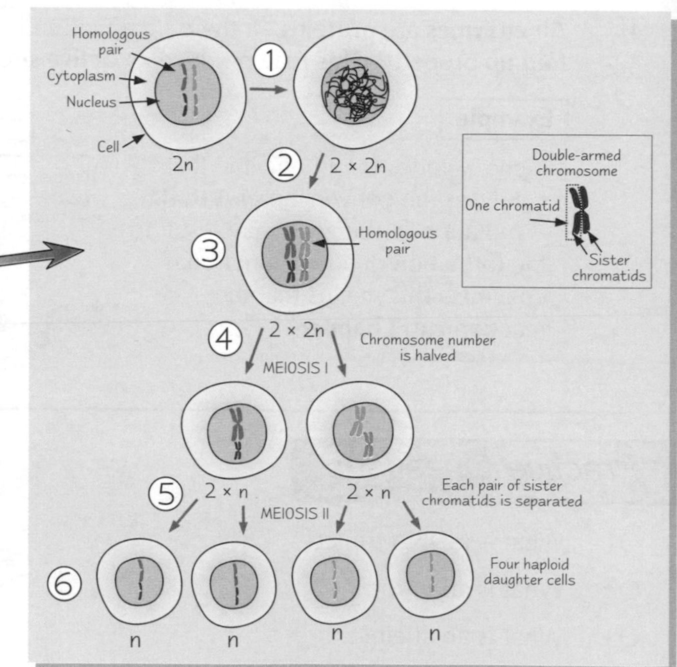


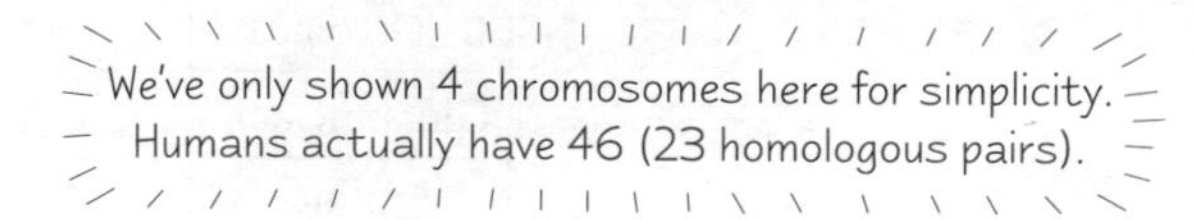


## Chromatids Cross Over in Meiosis I

During meiosis I, **homologous pairs** of chromosomes come together and pair up. The chromatids twist around each other and bits of **chromatids** swap over. The chromatids still contain the **same genes** but now have a different combination of **alleles**.

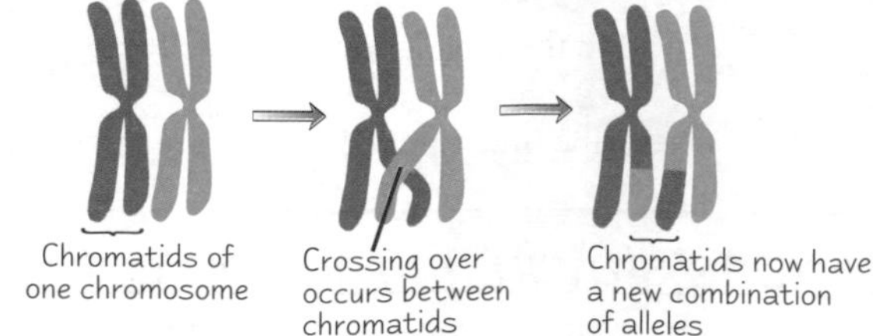

# Meiosis and Genetic Variation

## **Meiosis** Produces Cells that are **Genetically Different**

There are two main events during meiosis that lead to **genetic variation**:

### 1) Crossing over of chromatids

The **crossing over** of chromatids in meiosis I means that each of the **four daughter cells** formed from meiosis contain chromatids with **different alleles**:

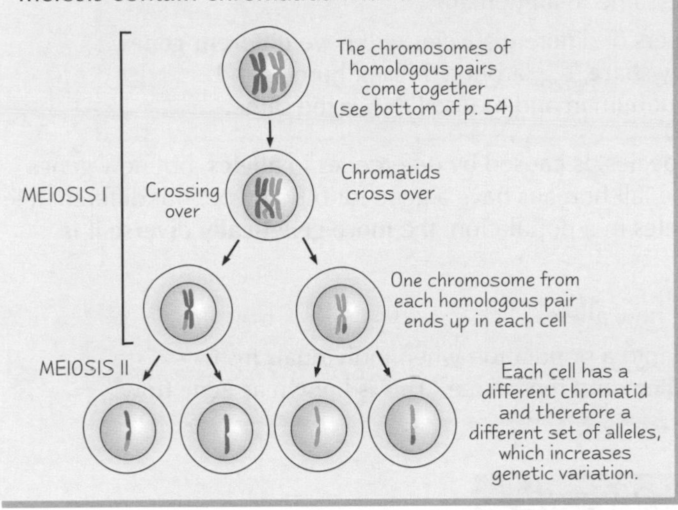


### 2) Independent segregation of chromosomes

1) The four daughter cells formed from meiosis have completely **different combinations of chromosomes.**
2) All your cells have a **combination** of chromosomes from your parents, half from your mum (**maternal**) and half from your dad (**paternal**).
3) When the gametes are produced, different **combinations** of those maternal and paternal **chromosomes** go into each cell.
4) This is called **independent segregation** (separation) of the chromosomes.

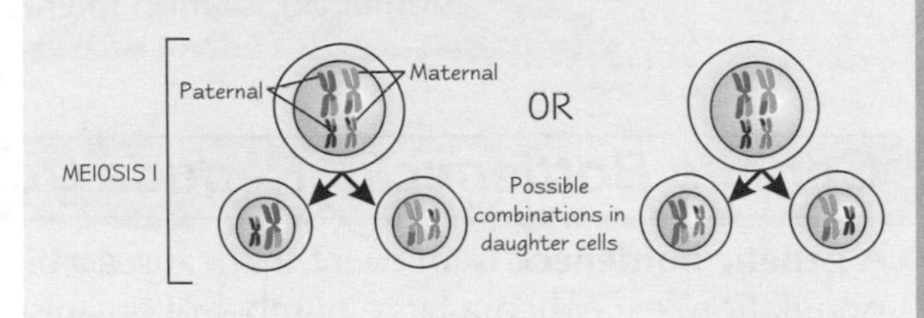


## Practice Questions

Q1 Explain what is meant by the terms haploid and diploid.

Q2 What happens to the chromosome number at fertilisation?

Q3 What is a chromatid?

Q4 How many divisions are there in meiosis?

Q5 How many cells does meiosis produce?

**Exam Questions**

Q1 Explain why it's important for gametes to have half the number of chromosomes as normal body cells. [2 marks]

Q2 Describe, using diagrams where appropriate, the process of meiosis. [6 marks]

Q3 a) Explain what crossing over is and how it leads to genetic variation. [4 marks]

b) Explain how independent segregation leads to genetic variation. [2 marks]

## Reproduction isn't as exciting as some people would have you believe...

*This page is quite tricky, so use the diagrams to help you understand — they might look evil, but they really do help. The key thing to understand is that meiosis produces four genetically different haploid (n) daughter cells. And that the genetic variation in the daughter cells occurs because of two processes — crossing over and independent segregation.*

# Genetic Diversity

*Genetic diversity describes the variety of alleles in a species or population.*

## *Variation in DNA can Lead to Genetic Diversity*

Genetic diversity is all about **variety**. The **more variety** in a population's DNA, the **more genetically diverse** it is.

1) Genetic diversity exists **within** a species. The DNA within a species varies **very little** though. **All** the members of the species will have the **same genes** but **different alleles**. For example, approximately 99.5% of DNA is the same in all humans.
2) The DNA of **different species** varies **a lot**. Members of different species will have **different genes**. The more **related** a species is, the more DNA they **share**, e.g. around 94% of human and chimpanzee DNA is the same, and around 85% of human and mouse DNA is the same.

Genetic diversity within a species, or a **population** of a species, is caused by differences in **alleles**, but new genes **don't appear** and old genes **don't disappear**. For example, all humans have a gene for blood type, but different alleles (versions) of blood type may come and go. The **more alleles** in a population, the **more genetically diverse** it is. Genetic diversity within a population is **increased** by:

1) **Mutations** in the DNA — forming new alleles.
2) Different alleles being introduced into a population when individuals from another population **migrate into them** and reproduce. This is known as **gene flow**.

## *Genetic Bottlenecks Reduce Genetic Diversity*

A **genetic bottleneck** is an event that causes a big **reduction** in a population, e.g. when a large number of organisms within a population **die** before reproducing. This reduces the number of **different alleles** in the gene pool and so reduces **genetic diversity**. The survivors **reproduce** and a larger population is created from a few individuals.

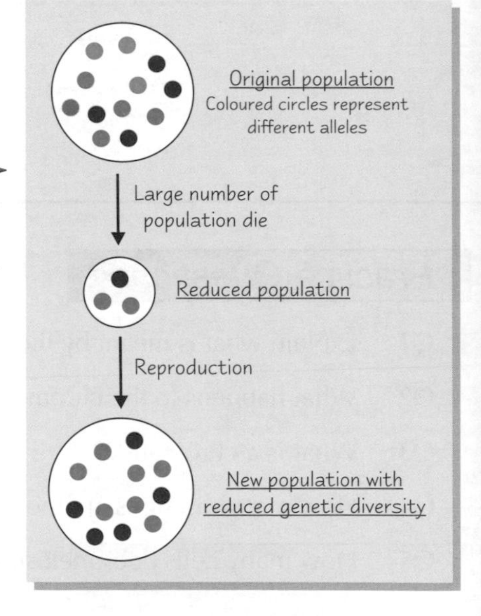


The gene pool is the complete range of alleles in a population.

**Example — Northern Elephant Seals**

**Northern elephant seals** were hunted by humans in the late 1800s. Their **original population** was reduced to around **50 seals** who have since produced a population of around 100 000. This new population has **very little** genetic diversity compared to the southern elephant seals who never suffered such a **reduction** in numbers.

Colin's offer to introduce new alleles into the population had yet to be accepted.

## *The Founder Effect is a Type of Genetic Bottleneck*

The **founder effect** describes what happens when just a **few** organisms from a population start a **new colony**. Only a small number of organisms have contributed their **alleles** to the **gene pool**. There's more **inbreeding** in the new population, which can lead to a **higher incidence** of genetic disease.

**Example — The Amish**

The **Amish population** of North America are all descended from a **small** number of Swiss who **migrated** there. The population shows **little genetic diversity**. They have remained **isolated** from the surrounding population due to their **religious beliefs**, so **few new alleles** have been introduced. The population suffers an unusually high incidence of certain **genetic disorders**.

The founder effect can occur as a result of **migration** leading to geographical **separation** or if a new colony is separated from the original population for **another reason**, such as **religion**.

# Genetic Diversity

## Selective Breeding Involves Choosing Which Organisms Reproduce

Changes in genetic diversity aren't just brought about by **natural events** like bottlenecks or migration. **Selective breeding** of plants and animals by humans has resulted in **reduced genetic diversity** in some populations. Selective breeding involves humans **selecting** which domesticated animals or strains of plants **reproduce** together in order to produce **high-yielding** breeds. For example:

1) A farmer wants a strain of **corn plant** that is tall and produces lots of ears, so he **breeds** a **tall** corn strain with one that produces **multiple ears**.
2) He selects the **offspring** that are tallest and have most ears, and breeds them **together**.
3) The farmer **continues** this until he produces a **very tall** strain that produces **multiple ears** of corn.

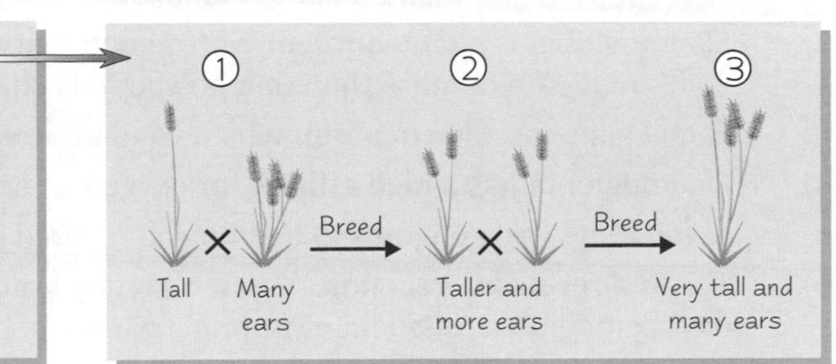


Selective breeding leads to a **reduction** in genetic diversity — once an organism with the **desired characteristics** (e.g. tall with multiple ears) has been produced, only that type of organism will continue being **bred**. So only similar organisms with **similar traits** and therefore **similar alleles** are bred together. It results in a type of **genetic bottleneck** as it reduces the **number of alleles** in the gene pool.

## Selective Breeding can Cause Problems for the Organisms Involved

You need to be able to discuss the **ethical issues** involved with selective breeding.

**Arguments FOR selective breeding**

1) It can produce **high-yielding** animals and plants.
2) It can be used to produce animals and plants that have increased **resistance** to disease. This means farmers have to use **fewer** drugs and pesticides.
3) Animals and plants could be bred to have increased tolerance of bad conditions, e.g. **drought** or **cold**.

**Arguments AGAINST selective breeding**

1) It can cause **health problems**. E.g. dairy cows are often **lame** and have a **short life expectancy** because of the extra strain making and carrying loads of milk puts on their bodies.
2) It **reduces genetic diversity**, which results in an increased incidence of **genetic disease** and an **increased susceptibility** to new diseases because of the lack of **alleles** in the population.

## Practice Questions

Q1 How does the founder effect reduce genetic diversity?

Q2 Describe the process of selective breeding.

Q3 Describe two arguments for and two arguments against selective breeding.

**Exam Questions**

Q1 Describe what a genetic bottleneck is and explain how it causes reduced genetic diversity within a population. [3 marks]

Q2 Describe what selective breeding is and explain why it leads to reduced genetic diversity within a population. [3 marks]

## *Sausage dogs didn't come from the wild...*

*You might think that selective breeding is a relatively new thing that we've developed with our knowledge of genetics... but you'd be wrong. We've been selectively breeding animals for yonks and yonks. All the different breeds of dog are just selectively bred strains which came from a general wolf-type dog back in the day. Even sausage dogs. Amazing...*

# Variation in Haemoglobin

*Haemoglobin's a protein that carries oxygen around the body. Different species have different versions of it depending on where each species lives. All of which adds up to two pages of no-holds-barred fun...*

## **Oxygen** is **Carried** Round the Body by **Haemoglobin**

1) **Red blood cells** contain **haemoglobin** (Hb).
2) Haemoglobin is a large **protein** with a **quaternary** structure (see p. 14 for more) — it's made up of **more than one** polypeptide chain (**four** of them in fact).
3) Each chain has a **haem group** which contains **iron** and gives haemoglobin its **red** colour.
4) Haemoglobin has a **high affinity for oxygen** — each molecule can carry **four oxygen molecules**.
5) In the lungs, oxygen **joins** to haemoglobin in red blood cells to form **oxyhaemoglobin**.
6) This is a **reversible reaction** — when oxygen leaves oxyhaemoglobin (**dissociates** from it) near the body cells, it turns back to haemoglobin.

$$Hb + 4O_2 \rightleftharpoons HbO_8$$

haemoglobin + oxygen ⇌ oxyhaemoglobin

'Affinity' for oxygen means tendency to combine with oxygen.

There are many **chemically similar** types of haemoglobin found in many different organisms, all of which carry out the **same function**. As well as being found in all vertebrates, haemoglobin is found in earthworms, starfish, some insects, some plants and even in some bacteria.

## **Haemoglobin Saturation** Depends on the **Partial Pressure** of **Oxygen**

1) The **partial pressure of oxygen** (**$pO_2$**) is a measure of **oxygen concentration.** The **greater** the concentration of dissolved oxygen in cells, the **higher** the partial pressure.
2) Similarly, the **partial pressure of carbon dioxide** (**$pCO_2$**) is a measure of the concentration of $CO_2$ in a cell.
3) Haemoglobin's **affinity** for oxygen **varies** depending on the **partial pressure** of **oxygen**:

> Oxygen **loads onto** haemoglobin to form oxyhaemoglobin where there's a **high $pO_2$**.
> Oxyhaemoglobin **unloads** its oxygen where there's a **lower $pO_2$**.

4) Oxygen enters blood capillaries at the **alveoli** in the **lungs**. Alveoli have a **high $pO_2$** so oxygen **loads onto** haemoglobin to form oxyhaemoglobin.
5) When **cells respire**, they use up oxygen — this **lowers the $pO_2$**. Red blood cells deliver oxyhaemoglobin to respiring tissues, where it unloads its oxygen.
6) The haemoglobin then returns to the lungs to pick up more oxygen.

There was no use pretending — the $pCH_4$ had just increased, and Keith knew who was to blame.

## **Dissociation Curves** Show How **Affinity for Oxygen** Varies

A **dissociation curve** shows how **saturated** the haemoglobin is with oxygen at any given partial pressure.

100% saturation means every haemoglobin molecule is carrying the maximum of 4 molecules of oxygen.

0% saturation means none of the haemoglobin molecules are carrying any oxygen.

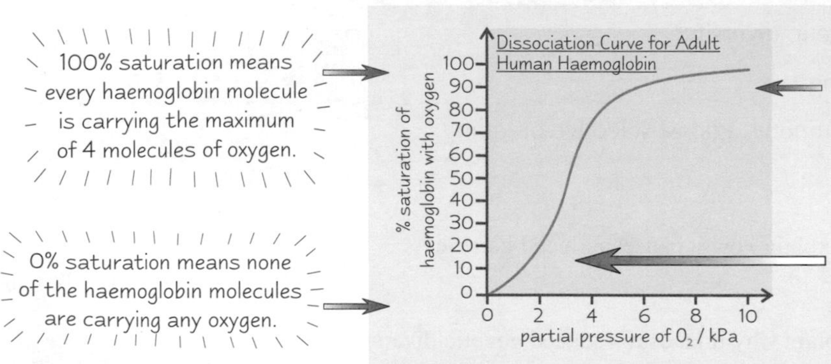


Where **$pO_2$ is high** (e.g. in the lungs), haemoglobin has a **high affinity** for oxygen (i.e. it will **readily combine** with oxygen), so it has a **high saturation** of oxygen.

Where **$pO_2$ is low** (e.g. in respiring tissues), haemoglobin has a **low affinity** for oxygen, which means it **releases oxygen** rather than combines with it. That's why it has a **low saturation** of oxygen.

The graph is '**S-shaped**' because when haemoglobin (Hb) combines with the **first $O_2$ molecule**, its **shape alters** in a way that makes it **easier** for other molecules to join too. But as the Hb starts to become saturated, it gets **harder** for more oxygen molecules to join. As a result, the curve has a **steep** bit in the middle where it's really easy for oxygen molecules to join, and **shallow** bits at each end where it's harder. When the curve is steep, a **small change in $pO_2$** causes a **big change** in the **amount of oxygen** carried by the Hb.

# Variation in Haemoglobin

## Carbon Dioxide Concentration Affects Oxygen Unloading

To complicate matters, haemoglobin gives up its oxygen **more readily** at **higher partial pressures of carbon dioxide** ($pCO_2$). It's a cunning way of getting more oxygen to cells during activity.

1) When cells respire they produce carbon dioxide, which **raises the $pCO_2$**.
2) This increases the rate of **oxygen unloading** — the dissociation curve '**shifts**' down. The saturation of blood with oxygen is **lower** for a given $pO_2$, meaning that **more oxygen** is being **released**.
3) This is called the **Bohr effect**.

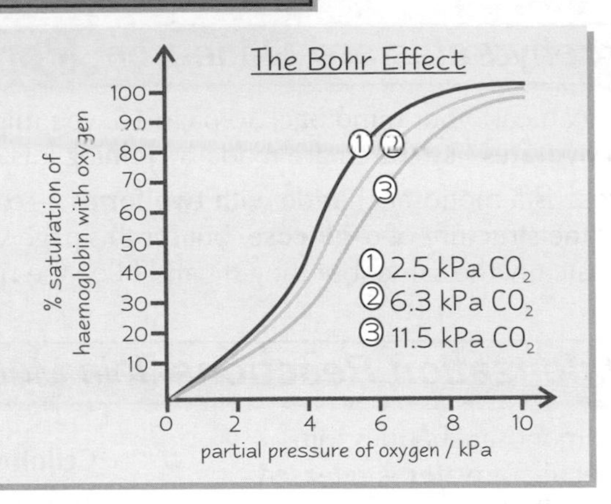


## Haemoglobin is Different in Different Organisms

Different organisms have different **types** of haemoglobin with different **oxygen transporting capacities**.

1) Organisms that live in environments with a **low concentration of oxygen** have haemoglobin with a **higher affinity** for oxygen than human haemoglobin — the dissociation curve is to the **left** of ours.
2) Organisms that are very **active** and have a **high oxygen demand** have haemoglobin with a **lower affinity** for oxygen than human haemoglobin — the curve is to the **right** of the human one.

A = animal living in depleted oxygen environment, e.g. a lugworm
B = animal living at high altitude where the partial pressure of oxygen is lower, e.g. a llama in the Andes.
C = human dissociation curve
D = active animal with a high respiratory rate living where there's plenty of available oxygen, e.g. a hawk.

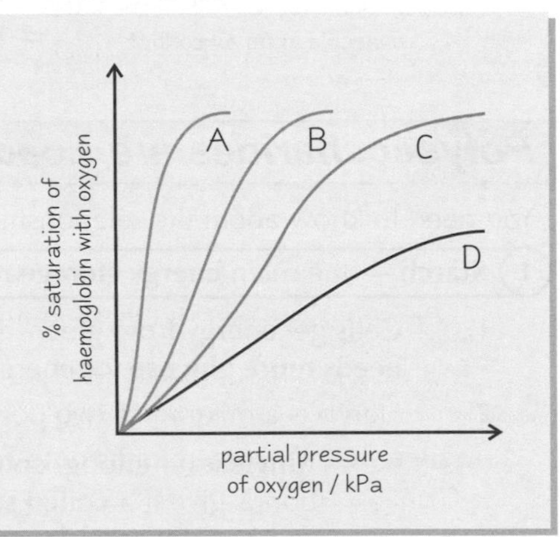


## Practice Questions

Q1 How many oxygen molecules can each haemoglobin molecule carry?

Q2 Where in the body would you find a low partial pressure of oxygen?

Q3 Why are oxygen dissociation curves S-shaped?

Q4 What is the Bohr effect?

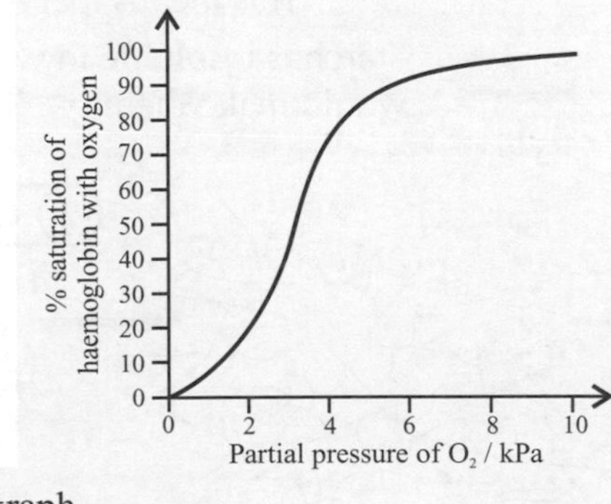


**Exam Questions**

Q1 The graph shows the oxygen dissociation curve for human haemoglobin. On the graph, sketch the curves you would expect for an earthworm (which lives in a low oxygen environment) and a human in a high carbon dioxide environment. Explain the position of your sketched curves. [6 marks]

Q2 Haemoglobin is a protein with a quaternary structure. Explain what this means. [1 mark]

## There's more than partial pressure on you to learn this stuff...

*Well, I don't know about you but after these two pages I need a sit-down. Most people get their knickers in a twist over partial pressure — it's not the easiest thing to get your head round. Whenever you see it written down just pretend it says concentration instead — cross it out and write concentration if you have to — and everything should become clearer. Honest.*

# Variation in Carbohydrates and Cell Structure

*Mmmmm, tasty tasty glucose... Unfortunately you have to learn its structure rather than eat it. You also need to cover the structure and function of tasty tasty starch and not-so-tasty-but-equally-important cellulose and glycogen. Enjoy...*

## Carbohydrates are Made from **Monosaccharides**

If you can cast your mind back to page 16, you might remember that **complex carbohydrates** like starch are made by **joining** lots of **monosaccharides** together.

Glucose is a monosaccharide with **two forms** — α and β. On page 16 you learnt about the structure of **α-glucose**, but for this page you need to know **β-glucose**... (It's basically the same, but the OH and H on the right are swapped around.)

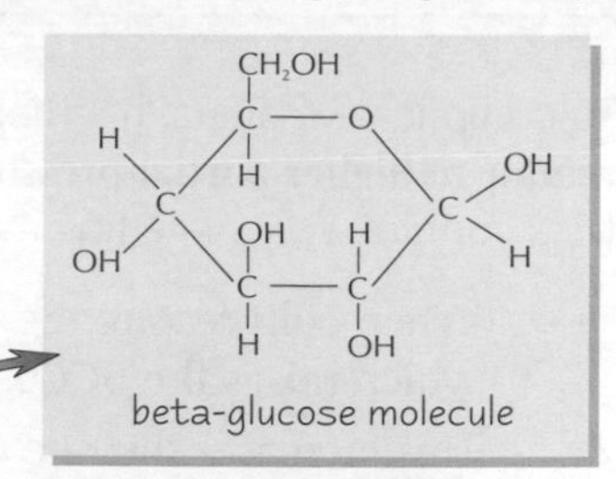

beta-glucose molecule

## **Condensation Reactions** Join Monosaccharides (Sugars) Together

When monosaccharides join, a molecule of **water** is **released**. This is called a **condensation reaction**. The bonds that join sugars together are called **glycosidic bonds**.

If you're asked to show a condensation reaction, don't forget to put the water molecule in as a product.

**Cellulose** is formed when beta-glucose is linked by condensation.

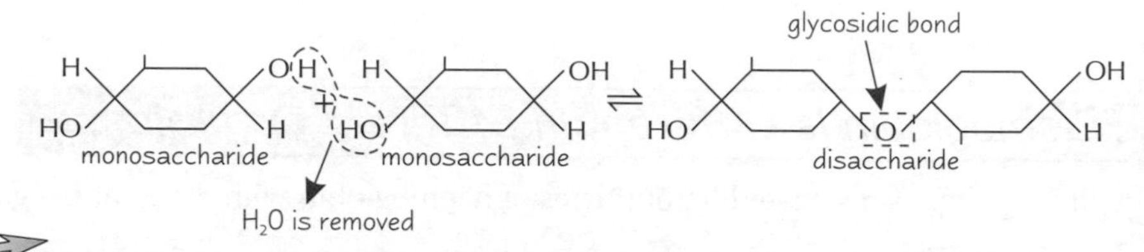


## **Polysaccharides** are **Loads of Sugars** Joined Together

You need to know about the relationship between the **structure** and **function** of three polysaccharides:

### 1 **Starch** — the main **energy storage material** in **plants**

1) Cells get **energy** from **glucose**. Plants **store** excess glucose as **starch** (when a plant **needs more glucose** for energy it **breaks down** starch to release the glucose).
2) Starch is a mixture of **two** polysaccharides of **alpha-glucose** — **amylose** and **amylopectin**:
   - **Amylose** — a long, **unbranched chain** of α–glucose. The angles of the glycosidic bonds give it a **coiled structure**, almost like a cylinder. This makes it **compact**, so it's really **good for storage** because you can **fit more in** to a small space.
   - **Amylopectin** — a long, **branched chain** of α–glucose. Its **side branches** allow the **enzymes** that break down the molecule to get at the **glycosidic bonds easily**. This means that the glucose can be **released quickly**.
3) Starch is **insoluble** in water so it doesn't cause water to enter cells by **osmosis**, which would make them swell (see p. 28). This makes it good for **storage**.

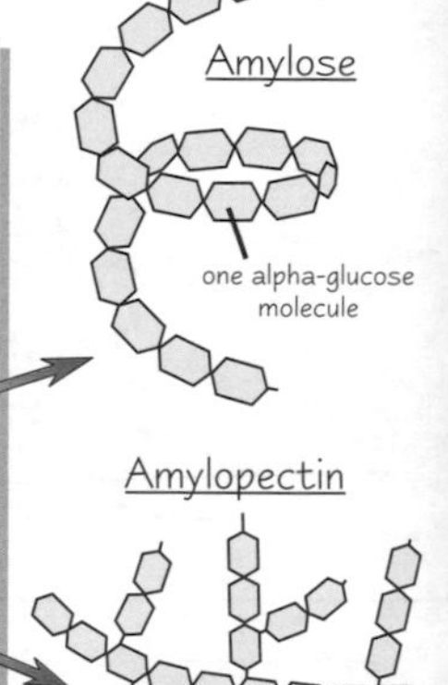


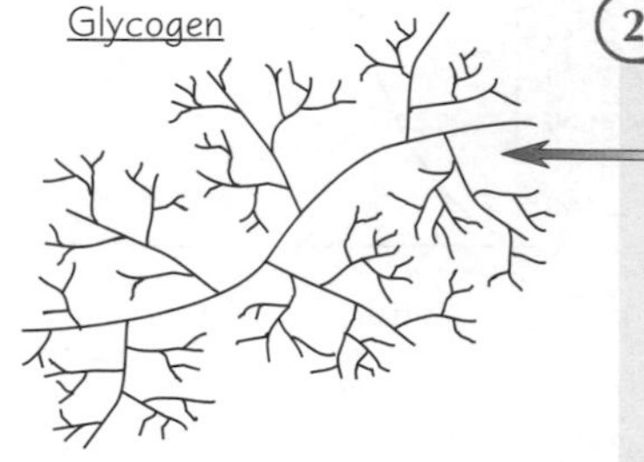


### 2 **Glycogen** — the main **energy storage material** in **animals**

1) Animal cells get **energy** from **glucose** too. But animals **store** excess glucose as **glycogen** — another polysaccharide of **alpha-glucose**.
2) Its structure is very similar to amylopectin, except that it has **loads** more **side branches** coming off it. Loads of branches means that stored glucose can be **released quickly**, which is **important for energy release** in animals.
3) It's also a very **compact** molecule, so it's good for storage.

### 3 **Cellulose** — the major component of **cell walls** in **plants**

1) Cellulose is made of **long, unbranched** chains of **beta-glucose**.
2) The **bonds** between the sugars are **straight**, so the cellulose chains are straight.
3) The cellulose chains are linked together by **hydrogen bonds** to form strong fibres called **microfibrils**. The strong fibres mean cellulose provides **structural support** for cells (e.g. in plant cell walls).

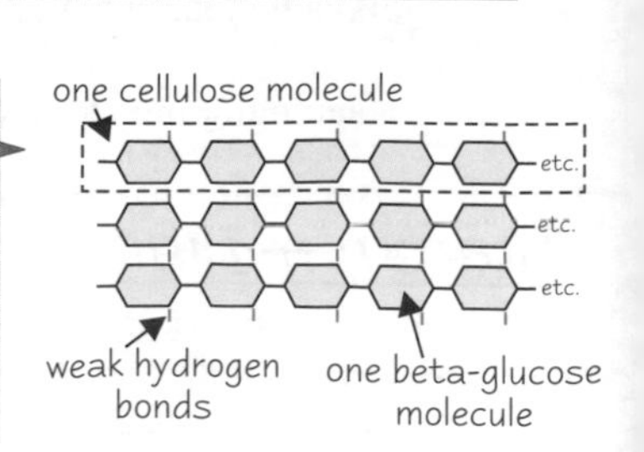

# Variation in Carbohydrates and Cell Structure

## ***Plant*** *and* ***Animal Cells*** *have* ***Similarities*** *and* ***Differences***

Way, way back in Unit 1 Section Three you learnt about animal cell structure and organelles. Well, you need to know about their structure for this section too so you can see how plant cell structure differs.

Most **animal** cells have the following parts — make sure you know them all:

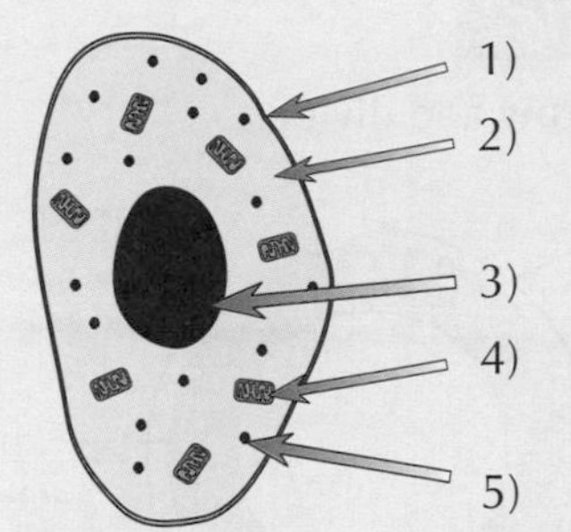

1) **Plasma membrane** — holds the cell together and controls what goes **in** and **out**.
2) **Cytoplasm** — gel-like substance where most of the **chemical reactions** happen. It contains **enzymes** (see page 18) that control these chemical reactions.
3) **Nucleus** — contains **genetic material** that controls the activities of the cell.
4) **Mitochondria** — where most of the reactions for **respiration** take place. Respiration releases **energy** that the cell needs to work.
5) **Ribosomes** — where **proteins** are made in the cell.

Plant cells usually have **all the bits** that **animal** cells have, plus a few **extra** things that animal cells **don't** have:

1) Rigid **cell wall** — made of **cellulose**. It **supports** and **strengthens** the cell.
2) **Permanent vacuole** — contains **cell sap**, a weak solution of sugar and salts.
3) **Chloroplasts** — where **photosynthesis** occurs, which makes food for the plant. They contain a **green** substance called **chlorophyll**.

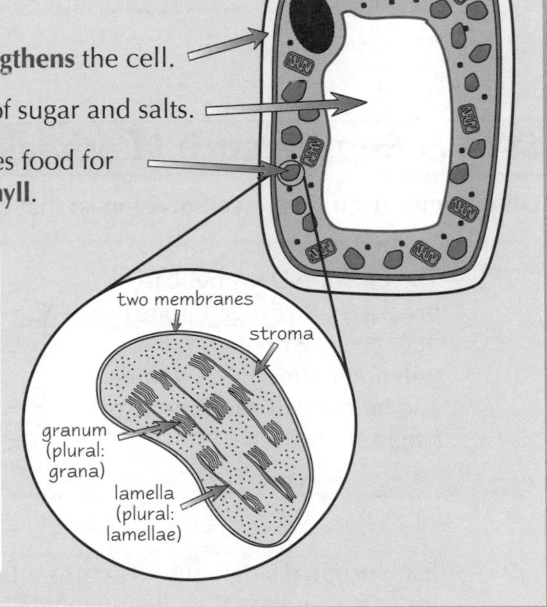


- Chloroplasts are surrounded by a **double membrane**, and also have membranes inside called **thylakoid membranes**. These membranes are stacked up in the chloroplast to form **grana**.
- Grana are linked together by **lamellae** — thin, flat pieces of thylakoid membrane.
- Some parts of photosynthesis happen in the **grana**, and other parts happen in the **stroma** (a thick fluid found in chloroplasts).

In the exam, you might get a question where you need to apply your **knowledge** of the **organelles** in a plant cell to explain why it's particularly **suited** to its **function**. Here are some tips:

- Think about **where** the cell's **located** in the plant — e.g. if it's exposed to **light**, then it'll have **lots** of **chloroplasts** to maximise **photosynthesis**.
- Think about **what** the cell **needs** to do its **job** — e.g. if the cell uses a lot of **energy**, it'll need lots of **mitochondria**. If it makes a lot of **proteins**, it'll need a lot of **ribosomes**.

## *Practice Questions*

Q1 Draw a diagram to illustrate a condensation reaction between two molecules of β-glucose.

Q2 Give three structures found in plant cells but not in animal cells.

**Exam Question**

Q1 Describe how the structure of starch makes it suited to its function. [6 marks]

## *Starch — I thought that was just for shirt collars...*

*It's important to understand that every cell in an organism is adapted to perform a function — you can always trace some of its features back to its function. Different cells even use the exact same molecules to do completely different things. Take glucose, for example — all plant cells use it to make cellulose, but they can also make starch from it if they need to store energy.*

# The Cell Cycle and DNA Replication

*Ever wondered how you grow from one tiny cell to a complete whole person? Or how that big cut you got in that horrific guitar strumming incident healed? No, oh... well you grow bigger and heal because your cells replicate and you need to learn the processes involved.*

## The Cell Cycle is the Process of Cell Growth and Division

The **cell cycle** is the process that all body cells from **multicellular organisms** use to **grow** and **divide**.

1) The cell cycle **starts** when a cell has been produced by cell division and **ends** with the cell dividing to produce two identical cells.
2) The cell cycle consists of a period of **cell growth** and **DNA replication**, called **interphase**, and a period of **cell division**, called **mitosis**.
3) Interphase (cell growth) is subdivided into three separate growth stages. These are called $G_1$, **S** and $G_2$.

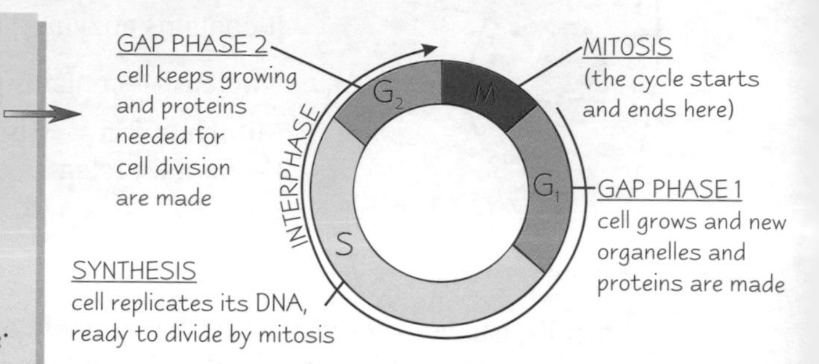


## DNA is Replicated in Interphase

**DNA copies itself** before **cell division** so that each new cell has the full amount of DNA.

1) The enzyme **DNA helicase breaks** the **hydrogen bonds** between the two **polynucleotide** DNA strands. The helix **unzips** to form two single strands.

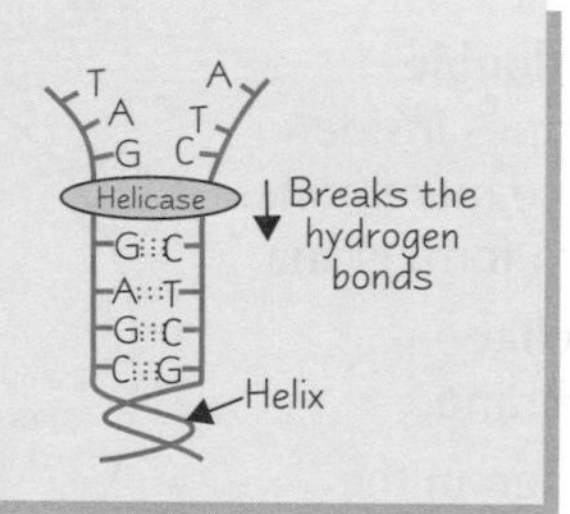


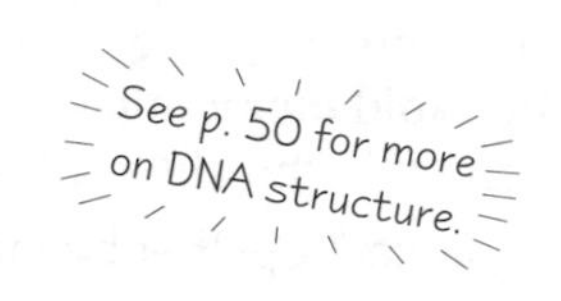

Mandy took her cells for a cycle.

2) Each **original** single strand acts as a **template** for a new strand. Free-floating DNA nucleotides join to the **exposed bases** on each original template strand by **specific base pairing** — A with T and C with G.

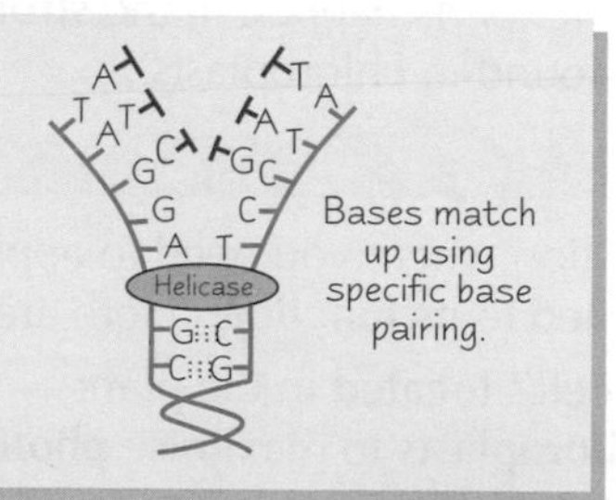


3) The nucleotides on the new strand are joined together by the enzyme **DNA polymerase**. Hydrogen bonds **form** between the bases on the original and new strand.

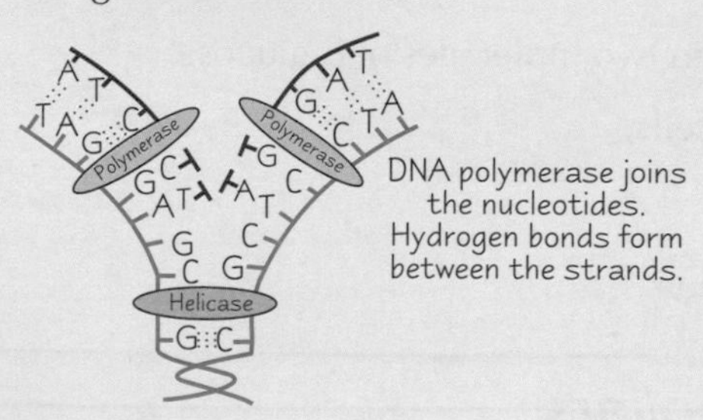


4) Each new DNA molecule contains **one strand** from the **original** DNA molecule and one **new strand.**

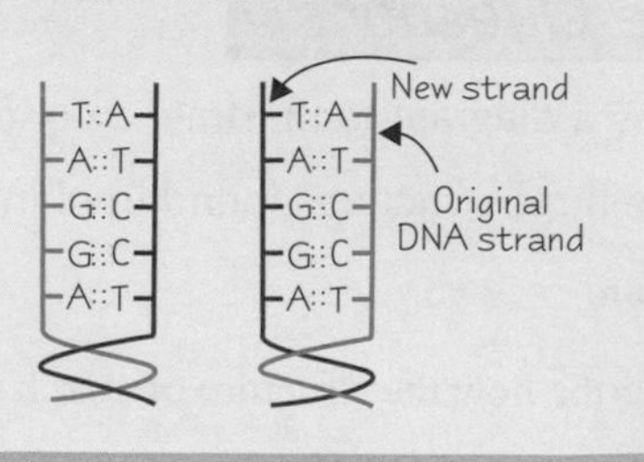


This type of copying is called **semi-conservative replication** because **half** of the new strands of DNA are from the **original** piece of DNA.

# The Cell Cycle and DNA Replication

## You May Have to *Interpret* Early *Experimental Work* About *DNA*

In the exam you might have to interpret **experimental evidence** that shows the **role** and **importance** of DNA. Here are some examples of the **early experiments** that were carried out. You **don't** need to learn them, just **understand** how the results show the role and importance of DNA.

### Evidence of hereditary molecules

An experiment with **mice** and two kinds of **pneumonia**, a **disease-causing** strain (**D**) and a **non-disease-causing** strain (**N**), showed there's a **hereditary molecule** (genetic material).

1) Mice injected with **strain D died** and with **strain N survived**.
2) **Killed D** was injected into mice — they **survived**.
3) **Killed D** and **live N** were injected together — they **died**.

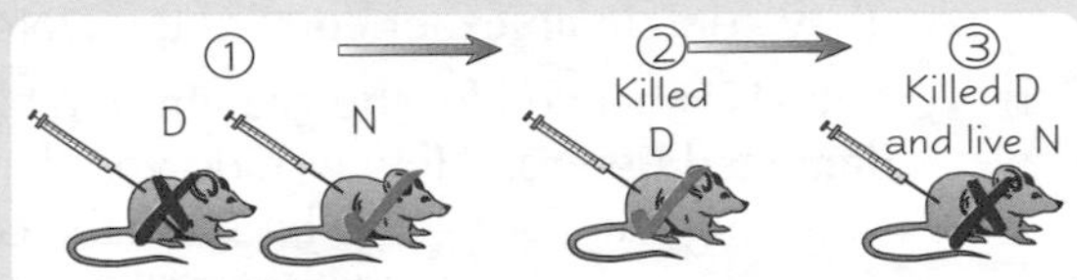


Killed D had **passed on** an inheritance molecule to the live N strain, making it **capable** of causing **disease**.

### Evidence that DNA is the genetic material

Scientists were unsure if the hereditary molecule was **DNA**, **RNA** or **protein**. They investigated it by treating the killed D strain with **protease** (destroys protein), **RNase** (destroys RNA) or **DNase** (destroys DNA) and then **injecting** it along with live N strain into mice. The strains that had been treated with DNase **didn't** kill the mice, so DNA was shown to be the **hereditary molecule** (genetic material).

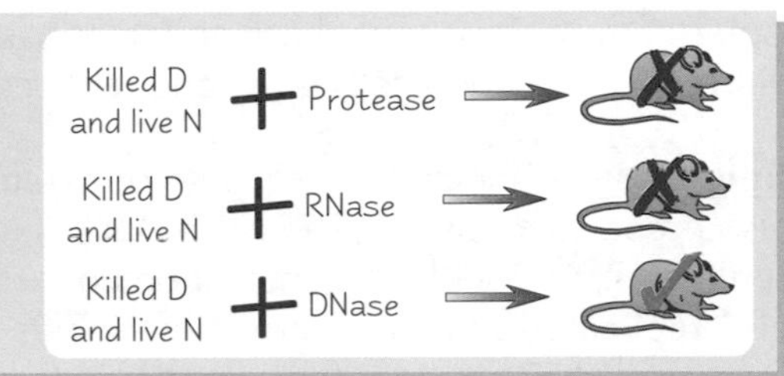


### More evidence that DNA is the genetic material

When viruses infect bacteria they **inject** their genetic material into the **cell**. So whatever viral material is found **inside** the bacterial cell must be the genetic material.

1) Scientists labelled the **DNA** of some viruses with radioactive **phosphate**, $^{32}$**P** (**blue**), and the **protein** of some more viruses with radioactive **sulfur**, $^{35}$**S** (**red**).
2) They then let the viruses **infect** some bacteria.
3) When they **separated** the bacteria and viruses they found $^{32}$**P** (**blue**) inside the bacteria and $^{35}$**S** (**red**) on the outside, providing **evidence** that DNA was the **genetic material**.

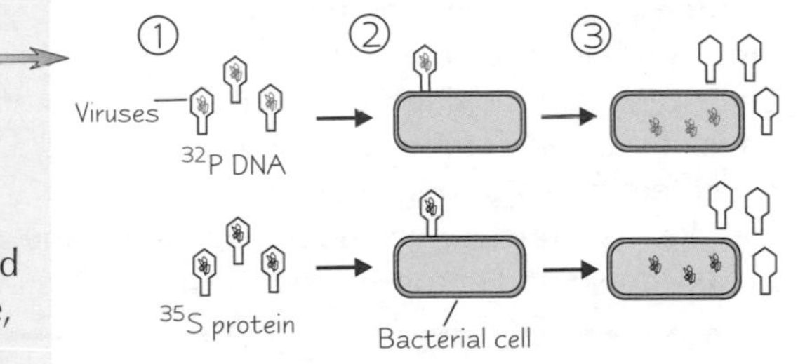


## Practice Questions

Q1 Name the two main stages of the cell cycle.

Q2 Why is DNA replication described as semi-conservative?

Q3 Name two enzymes involved in DNA replication.

**Exam Question**

Q1 a) Fill in the missing base pairs on the diagram opposite. [1 mark]

A A G C C T
T       G

b) Draw a diagram to show the DNA molecule on the right after it has replicated. Label the original and new strands. [2 marks]

## *I went through a gap phase — I just love their denim...*

*DNA and its self-replication is important — so make sure you understand what's going on. Diagrams are handy for learning stuff like this. I don't just put them in to keep myself amused you know — so get drawing and learning.*

# Cell Division — Mitosis

*I don't like cell division. There, I've said it. It's unfair of me, because if it wasn't for cell division I'd still only be one cell big. It's all those diagrams that look like worms nailed to bits of string that put me off.*

## Mitosis is Cell Division that Produces Genetically Identical Cells

1) There are two types of cell division — **mitosis** and **meiosis** (see p. 54 for more on meiosis).
2) Mitosis is the form of cell division that occurs during the **cell cycle**.
3) In **mitosis** a **parent cell** divides to produce **two genetically identical daughter cells** (they contain an **exact copy** of the **DNA** of the parent cell).
4) Mitosis is needed for the **growth** of multicellular organisms (like us) and for **repairing damaged tissues**. How else do you think you get from being a baby to being a big, strapping teenager — it's because the cells in our bodies grow and divide.

## Mitosis has Four Division Stages

Mitosis is really one **continuous process**, but it's described as a series of **division stages** — prophase, metaphase, anaphase and telophase. **Interphase** comes **before** mitosis in the cell cycle — it's when cells grow and replicate their DNA ready for division (see p. 62).

<u>Interphase</u> — The cell carries out normal functions, but also prepares to divide. The cell's **DNA** is unravelled and **replicated**, to double its genetic content. The **organelles** are also **replicated** so it has spare ones, and its ATP content is increased (ATP provides the energy needed for cell division).

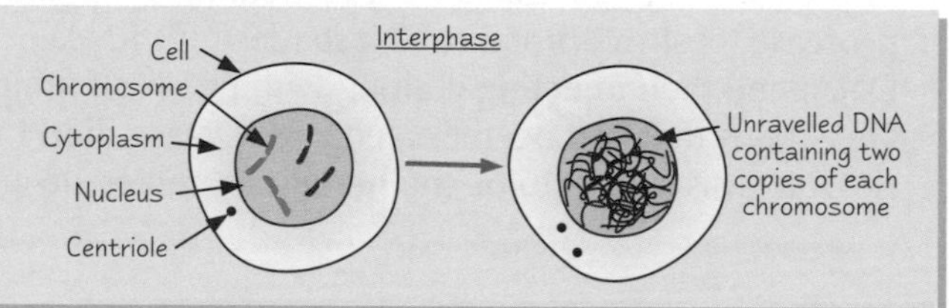


1) <u>Prophase</u> — The **chromosomes condense**, getting shorter and fatter. Tiny bundles of protein called **centrioles** start moving to opposite ends of the cell, forming a network of protein fibres across it called the **spindle**. The **nuclear envelope** (the membrane around the nucleus) **breaks down** and chromosomes lie free in the cytoplasm.



As mitosis begins, the chromosomes are made of two strands joined in the middle by a <u>centromere</u>.
The separate strands are called <u>chromatids</u>.

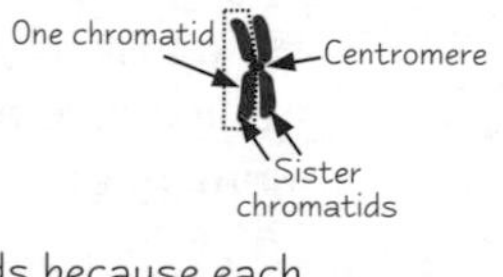


There are two strands because each chromosome has already made an <u>identical copy</u> of itself during <u>interphase</u>. When mitosis is over, the chromatids end up as one-strand chromosomes in the new daughter cells.

2) <u>Metaphase</u> — The chromosomes (each with two chromatids) **line up** along the middle of the cell and become **attached** to the **spindle** by their **centromere**.

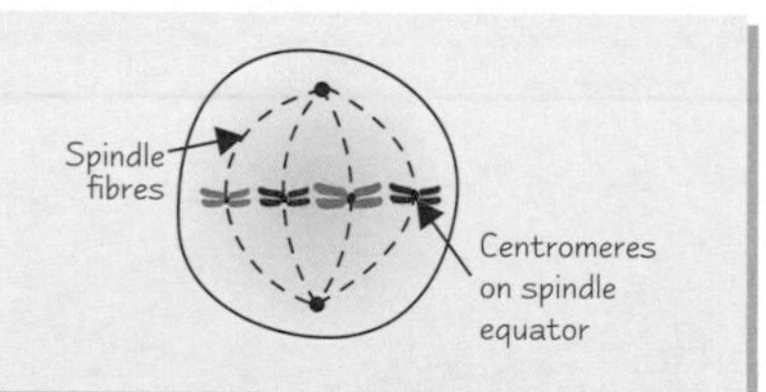


You need to be able to recognise each stage in mitosis from <u>photographs</u>. But don't worry — they'll look pretty much like the diagrams here.

3) <u>Anaphase</u> — The centromeres divide, **separating** each pair of sister **chromatids**. The spindles contract, pulling chromatids to opposite ends of the cell, centromere first.

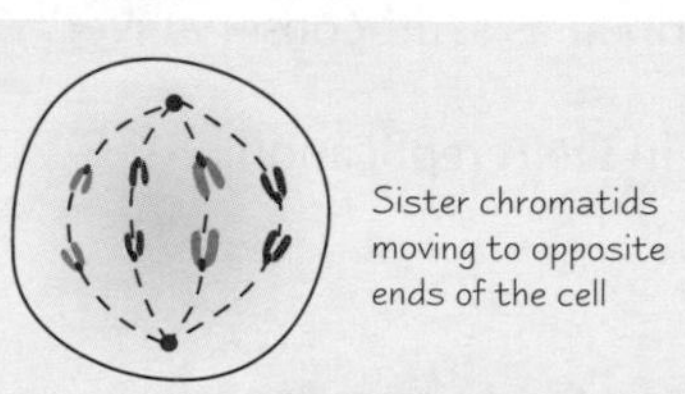


4) <u>Telophase</u> — The chromatids reach the **opposite poles** on the spindle. They uncoil and become long and thin again. They're now called **chromosomes** again. A **nuclear envelope** forms around each group of chromosomes, so there are now **two nuclei**. The **cytoplasm divides** and there are now **two daughter cells** that are **genetically identical** to the original cell and to each other. Mitosis is finished and each daughter cell starts the **interphase** part of the cell cycle to get ready for the next round of mitosis.

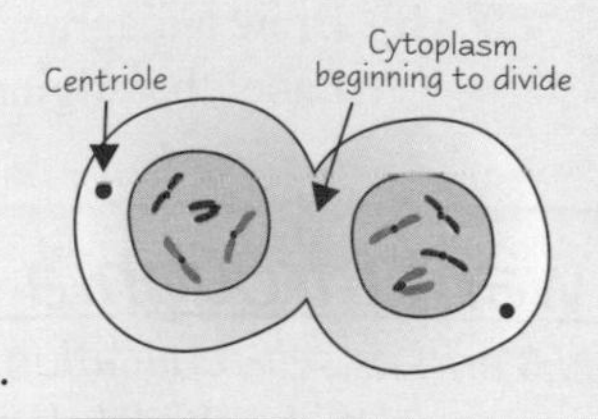

# Cell Division — Mitosis

## **Cancer** is the Result of **Uncontrolled Cell Division**

*Mutations are changes in the base sequence of an organism's DNA* *(see p. 53)**.*

1) Cell growth and cell division are **controlled by genes**.
2) Normally, when cells have divided enough times to make **enough new cells**, they stop. But if there's a **mutation** in a gene that controls cell division, the cells can **grow out of control**.
3) The cells **keep on dividing** to make more and more cells, which form a **tumour**.
4) **Cancer** is a tumour that **invades** surrounding tissue.

## Some **Cancer Treatments** Target the **Cell Cycle**

Some treatments for cancer are designed to **disrupt** the cell cycle.
These treatments don't **distinguish** tumour cells from normal cells though — they also **kill normal body cells** that are dividing. However, tumour cells **divide much more frequently** than normal cells, so the treatments are **more likely** to kill tumour cells. Some cell cycle **targets** of cancer treatments include:

1) **G1 (cell growth and protein production)** — Some chemical drugs (chemotherapy) prevent the **synthesis of enzymes** needed for DNA replication. If these aren't produced, the cell is unable to enter the **synthesis phase** (S), disrupting the cell cycle and forcing the cell to **kill itself**.
2) **S phase (DNA replication)** — **Radiation** and some drugs **damage DNA**. When the cell gets to S phase it checks for **damaged DNA** and if any is detected it **kills itself**, preventing **further** tumour growth.

Because cancer treatments **kill normal cells** too certain steps are taken to **reduce the impact** on normal body cells:

1) A **chunk** of tumour is often removed first using **surgery**. This removes a lot of tumour cells and increases the access of any left to nutrients and oxygen, which triggers them to enter the **cell cycle**, making them **more susceptible** to treatment.
2) **Repeated treatments** are given with periods of **non-treatment** (breaks) in between. A **large dose** could kill **all the tumour** but also so many normal cells that the patient could **die**. Repeated treatments with **breaks** allows the body to **recover** and produce new cells. The treatment is **repeated** as any tumour cells **not killed** by the treatment will keep **dividing and growing** during the breaks too. The break period is kept short so the body can **recover** but the **cancer** can't grow back to the same size as before.

## Practice Questions

Q1 Give the two main uses of mitosis.

Q2 List the four stages of mitosis.

Q3 Describe how tumours are formed.

Q4 Describe how repeated cancer treatment with breaks can help to reduce the impact on normal body cells.

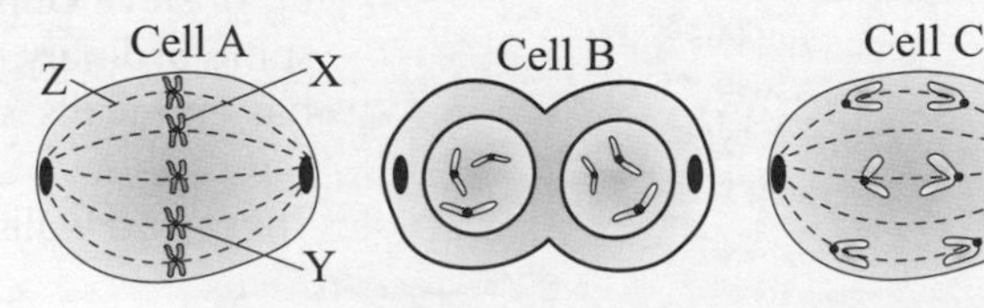


**Exam Question**

Q1 The diagrams show cells at different stages of mitosis.

a) For each of the cells A, B and C state the stage of mitosis, giving a reason for your answer. [6 marks]

b) Name the structures labelled X, Y and Z in cell A. [3 marks]

## Doctor, I'm getting short and fat — don't worry, it's just a phase...

*Quite a lot to learn on these pages — but it's all important stuff, so no slacking. Mitosis is vital — it's how cells multiply and how organisms like us grow. Don't forget — the best way to learn is to get drawing those diagrams.*

# Cell Differentiation and Organisation

*In complex multicellular organisms like me (well, I wouldn't say I'm complex, but multicellular at least), cells are adapted for different jobs. And all these cells are organised to work together.*

## Cells of Multicellular Organisms Can Differentiate

1) **Multicellular organisms** are made up from many **different** cell types, e.g. nerve cells, muscle cells, white blood cells.
2) **All** these cell types are **specialised** — they're designed to carry out **specific functions** (see below).
3) The **structure** of each specialised cell type is **adapted** to suit its particular job.
4) The **process** of **becoming specialised** is called **differentiation.**

Joe knew his cells were specialised — specialised to look good.

## Differentiated Cells are Adapted for Specific Functions

Here are two examples of differentiated cells to show you how they're adapted for their function:

1) **Squamous epithelium cells** are found in many places. They're **thin**, with not much cytoplasm. In the lungs they line the **alveoli** and are thin to allow gases to pass through them easily.

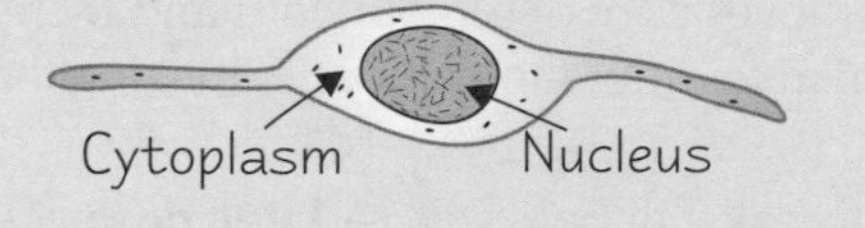


2) **Palisade mesophyll cells** in leaves are where **photosynthesis** occurs. They contain **many chloroplasts**, so they can absorb as much sunlight as possible. The walls are **thin**, so carbon dioxide can **easily enter**.

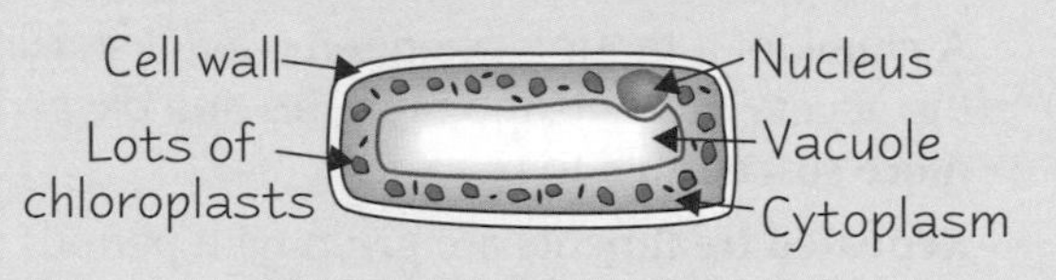


## Similar Cells are Organised into Tissues

Similar cells are grouped together into **tissues**. Here are some examples:

1) **Squamous epithelium tissue** is a **single layer** of **flat cells** lining a surface. Squamous epithelium tissue is found in many places including the alveoli in the lungs.

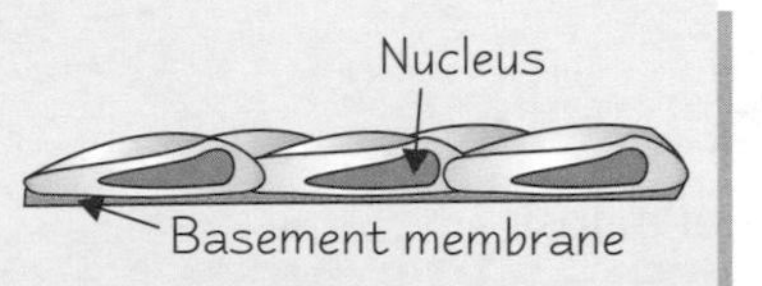


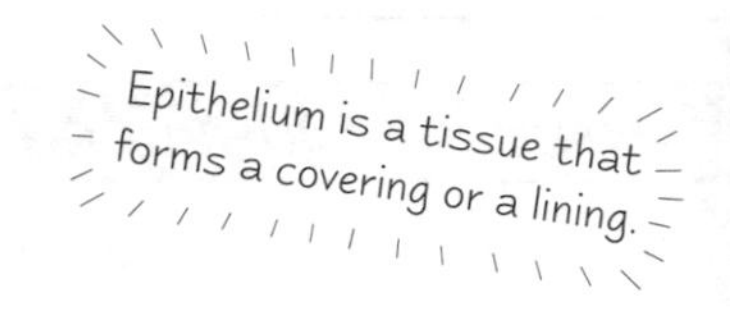


Tissues aren't always made up of one type of cell. Some tissues include different types of cell working together.

2) **Phloem tissue** **transports sugars** around the plant. It's arranged in **tubes** and is made up of **sieve cells**, **companion cells**, and some **ordinary** plant cells. Each sieve cell has end walls with **holes** in them, so that sap can move easily through them. These end walls are called **sieve plates**.

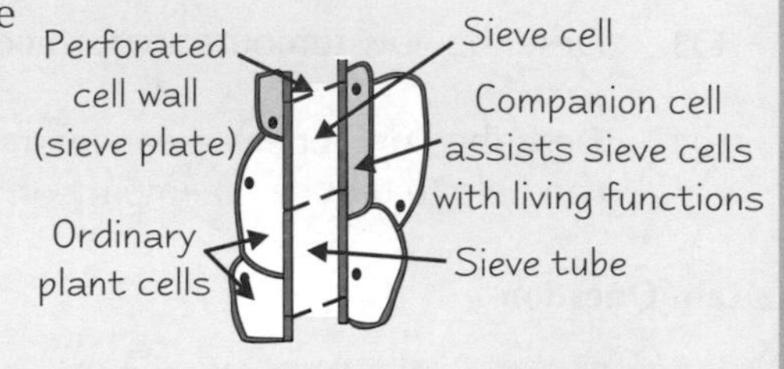


3) **Xylem tissue** is a plant tissue with two jobs — it **transports water** around the plant, and it **supports** the plant. It contains **xylem vessel cells** and **parenchyma cells**.

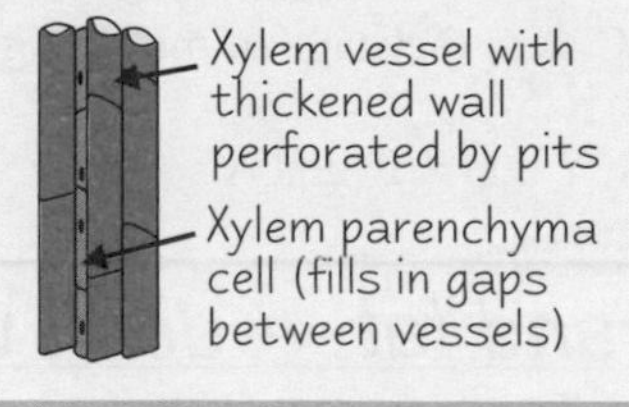

# Cell Differentiation and Organisation

## *Tissues* are Organised into *Organs*

An **organ** is a group of different tissues that **work together** to perform a particular function. Here are two examples:

The leaf is an example of a plant organ. It's made up of the following **tissues**:

1) **Lower epidermis** — contains stomata (pores) to let air in and out for gas exchange.
2) **Spongy mesophyll** — full of spaces to let gases circulate.
3) **Palisade mesophyll** — most photosynthesis occurs here.
4) **Xylem** — carries water to the leaf.
5) **Phloem** — carries sugars away from the leaf.
6) **Upper epidermis** — covered in a waterproof waxy cuticle to reduce water loss.

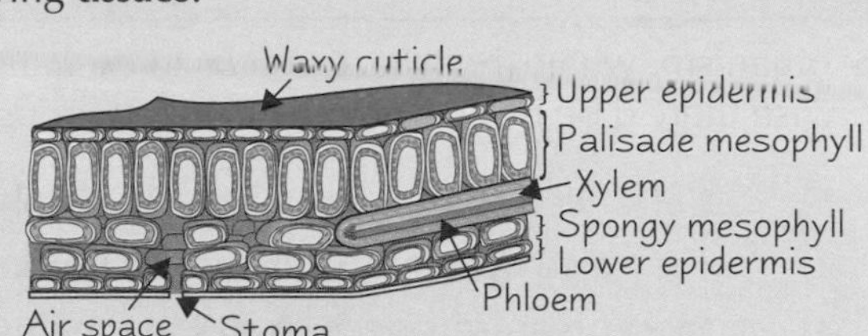


The lungs are an example of an animal organ. They're made up of the following **tissues**:

1) **Squamous epithelium tissue** — surrounds the alveoli (where gas exchange occurs).
2) **Fibrous connective tissue** — forms a continuous mesh around the lungs and contains fibres that help to force the air back out of the lungs when exhaling.
3) **Blood vessels** — capillaries surround the alveoli.

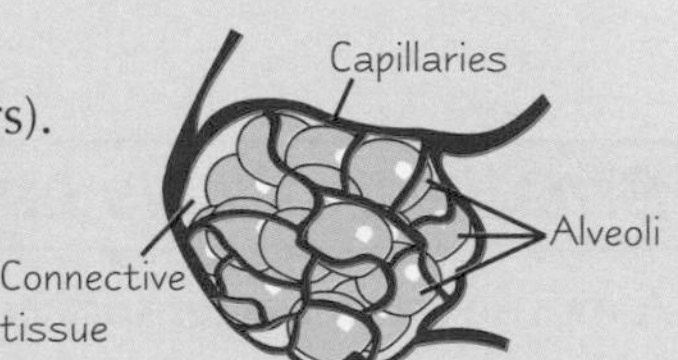


## *Organs* are Organised into *Systems*

Organs work together to form **organ systems** — each system has a **particular function**. Here are some examples:

The Circulatory System allows the transport of gases and other substances around the body. It includes:

1) **The heart** — pumps the blood around the body.
2) **Blood vessels** — carries the blood to the tissues.

The Shoot System in plants includes:
**leaves** (site of photosynthesis), **buds** (growing regions), **stems** (for support) and **flowers** (for sexual reproduction).

The Respiratory System brings oxygen into the body and removes carbon dioxide. It includes:

1) **The lungs** — where gas exchange occurs.
2) **The trachea** — allows air flow.
3) **The bronchi** — carry the air into the lungs.

## Practice Questions

Q1 What is meant by cell differentiation?

Q2 Define what is meant by a tissue.

Q3 Give one animal and one plant example of an organ system.

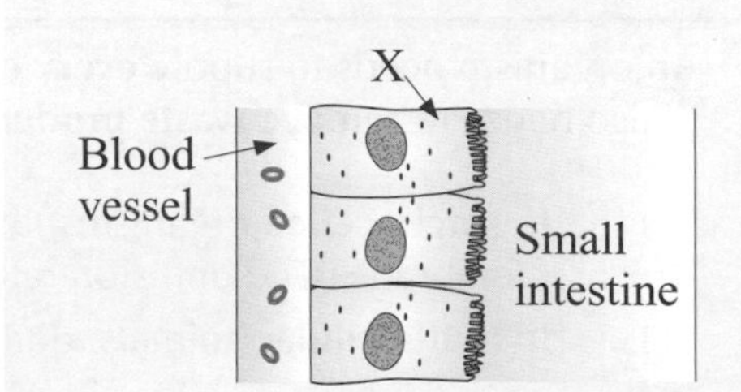


**Exam Questions**

Q1 Tissue X above is found lining the small intestine, where nutrients are absorbed into the bloodstream. Outline how it is adapted for its function. [4 marks]

Q2 The liver is made of hepatocyte cells that form the main tissue, blood vessels to provide nutrients and oxygen, and connective tissue that holds the organ together. Discuss whether the liver is best described as a tissue or an organ. [2 marks]

## *Soft and quilted — the best kind of tissue...*

*The important thing to remember from these pages is that in multicellular organisms, cells are adapted for their function. You don't need to learn the examples off by heart — just understand how the adaptations are related to the function.*

# Size and Surface Area

*Exchanging things with the environment is pretty easy if you're a single-celled organism, but if you're multicellular it all gets a bit more complicated... and it's all down to this 'surface area to volume ratio' malarkey.*

## Organisms Need to **Exchange Substances** with their **Environment**

Every organism, whatever its size, needs to exchange things with its environment. Otherwise there'd be no such thing as poop scoops...

1) Cells need to take in **oxygen** (for aerobic respiration) and **nutrients**.
2) They also need to excrete **waste products** like **carbon dioxide** and **urea**.
3) Most organisms need to stay at roughly the **same temperature**, so **heat** needs to be exchanged too.

Raj was glad he'd exchanged his canoe for a bigger boat.

How easy the exchange of substances is depends on the organism's **surface area to volume ratio**.

## **Smaller** Animals have **Higher Surface Area : Volume Ratios**

A mouse has a bigger surface area **relative to its volume** than a hippo. This can be hard to imagine, but you can prove it mathematically. Imagine these animals as cubes:

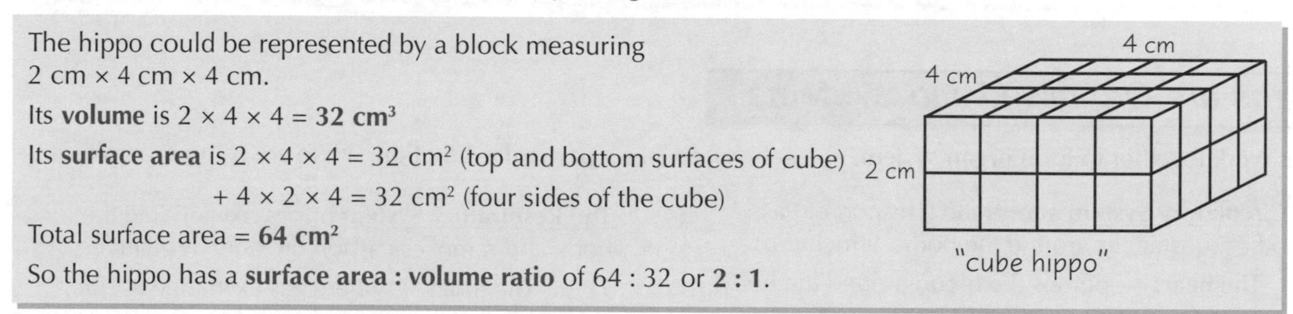


The hippo could be represented by a block measuring 2 cm × 4 cm × 4 cm.

Its **volume** is 2 × 4 × 4 = **32 cm³**

Its **surface area** is 2 × 4 × 4 = 32 cm² (top and bottom surfaces of cube)
+ 4 × 2 × 4 = 32 cm² (four sides of the cube)

Total surface area = **64 cm²**

So the hippo has a **surface area : volume ratio** of 64 : 32 or **2 : 1**.

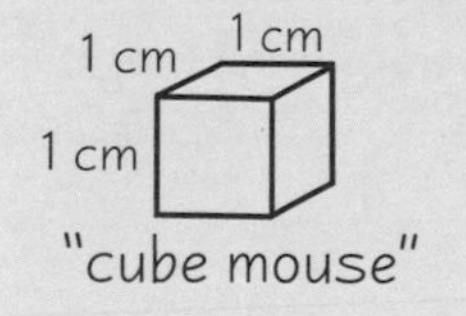


Compare this to a cube mouse measuring 1 cm × 1 cm × 1 cm.

Its **volume** is 1 x 1 x 1 = **1 cm³**

Its **surface area** is 6 x 1 x 1 = **6 cm²**

So the mouse has a **surface area : volume ratio** of 6 : 1

The cube mouse's surface area is six times its volume, but the cube hippo's surface area is only twice its volume. Smaller animals have a bigger surface area compared to their volume.

## **Multicellular** Organisms need **Exchange Organs** and **Mass Transport Systems**

An organism needs to supply **every one of its cells** with substances like **glucose** and **oxygen** (for respiration). It also needs to **remove waste products** from every cell to avoid damaging itself.

1) In **single-celled** organisms, these substances can **diffuse directly** into (or out of) the cell across the cell surface membrane. The diffusion rate is quick because of the small distances the substances have to travel (see p. 28).
2) In **multicellular** animals, diffusion across the outer membrane is **too slow**, for two reasons:
   - Some cells are **deep within the body** — there's a big distance between them and the **outside environment**.
   - Larger animals have a **low surface area to volume ratio** — it's difficult to exchange **enough** substances to supply a **large volume of animal** through a relatively **small outer surface**.

So rather than using straightforward diffusion to absorb and excrete substances, multicellular animals need specialised **exchange organs** (like lungs — see p. 35).

They also need an efficient system to carry substances to and from their individual cells — this is **mass transport**. In mammals, 'mass transport' normally refers to the **circulatory system** (see p. 40), which uses **blood** to carry glucose and oxygen around the body. It also carries **hormones**, **antibodies** (p. 6) and **waste** like $CO_2$.

# Size and Surface Area

## **Body Size** and **Shape** Affect **Heat Exchange**

As well as creating **waste products** that need to be transported away, the metabolic activity inside cells creates **heat**. Staying at the right temperature is difficult, and it's pretty heavily influenced by your **size** and **shape**...

**Size**

The **rate of heat loss** from an organism depends on its **surface area**. As you saw on the previous page, if an organism has a large volume, e.g. a hippo, its surface area is relatively **small**. This makes it **harder** for it to lose heat from its body. If an organism is small, e.g. a mouse, its relative surface area is **large**, so heat is lost more **easily**.

**Shape**

1) Animals with a **compact** shape have a **small surface area** relative to their volume — **minimising heat loss** from their surface.
2) Animals with a **less compact** shape (those that are a bit **gangly** or have **sticky outy** bits) have a **larger surface area** relative to their volume — this **increases heat loss** from their surface.
3) Whether an animal is compact or not depends on the **temperature** of its **environment**. Here's an example:

Arctic fox
Body temperature 37 °C
Average outside temperature 0 °C

The Arctic fox has **small ears** and a **round head** to **reduce** its SA : V ratio and heat loss.

African bat-eared fox
Body temperature 37 °C
Average outside temperature 25 °C

The African bat-eared fox has **large ears** and a more **pointed nose** to **increase** its SA : V ratio and heat loss.

European fox
Body temperature 37 °C
Average outside temperature 12 °C

The European fox is **intermediate** between the two, matching the temperature of its environment.

## Organisms have **Behavioural** and **Physiological Adaptations** to Aid Exchange

Not all organisms have a body size or shape to suit their climate — some have **other adaptations** instead...

1) Animals with a high SA : volume ratio tend to **lose more water** as it evaporates from their surface. Some **small desert mammals** have **kidney structure adaptations** so that they produce **less urine** to compensate.
2) **Smaller animals** living in **colder regions** often have a much **higher metabolic rate** to compensate for their high SA : volume ratio — this helps to keep them warm by creating **more heat**. To do this they need to eat large amounts of **high energy foods** such as seeds and nuts.
3) Smaller mammals may have thick layers of **fur** or **hibernate** when the weather gets really cold.
4) **Larger organisms** living in **hot regions**, such as elephants and hippos, find it hard to keep cool as their heat loss is relatively slow. **Elephants** have developed **large flat ears** which **increase** their **surface area**, allowing them to lose more heat. **Hippos** spend much of the day in the **water** — a **behavioural adaptation** to help them lose heat.

## Practice Questions

Q1 Give four things that organisms need to exchange with their environment.

Q2 Describe how body shape affects heat exchange.

**Exam Question**

Q1 Explain why diffusion is not an efficient transport system for large mammals. [3 marks]

## Cube animals indeed — it's all gone a bit Picasso...

*You need to understand why single-celled organisms and large multicellular organisms use different methods for exchange. Most multicellular organisms couldn't survive using diffusion alone — that's why they have exchange organs.*

# Gas Exchange

*Lots of organisms have developed adaptations to improve their rate of gas exchange. It's a tricky business if you're an insect or a plant though — you've got to exchange enough gas but avoid losing all your water and drying to a crisp...*

## Gas Exchange Surfaces have **Two** Major **Adaptations**

Most gas exchange surfaces have two things in common:

1) They have a **large surface area**.
2) They're **thin** (often just one layer of epithelial cells) — this provides a **short diffusion pathway** across the gas exchange surface.

The organism also maintains a **steep concentration gradient** of gases across the exchange surface.

All these features **increase** the **rate of diffusion.**

## **Single-celled Organisms** Exchange Gases across their **Body Surface**

1) Single-celled organisms absorb and release gases by **diffusion** through their **outer surface.**
2) They have a relatively **large surface area**, a **thin surface** and a **short diffusion pathway** (oxygen can take part in **biochemical reactions** as soon as it **diffuses** into the cell) — so there's **no need** for a gas exchange system.

## **Fish** Use a **Counter-Current System** for Gas Exchange

There's a **lower concentration** of oxygen in water than in air. So **fish** have special **adaptations** to get enough of it.

1) Water, containing oxygen, enters the fish through its **mouth** and passes out through the gills.
2) Each gill is made of lots of **thin plates** called **gill filaments**, which give a **big surface area** for **exchange** of **gases**.
3) The gill filaments are covered in lots of tiny structures called **lamellae**, which **increase** the **surface area** even more.
4) The lamellae have lots of **blood capillaries** and a thin surface layer of cells to speed up diffusion.

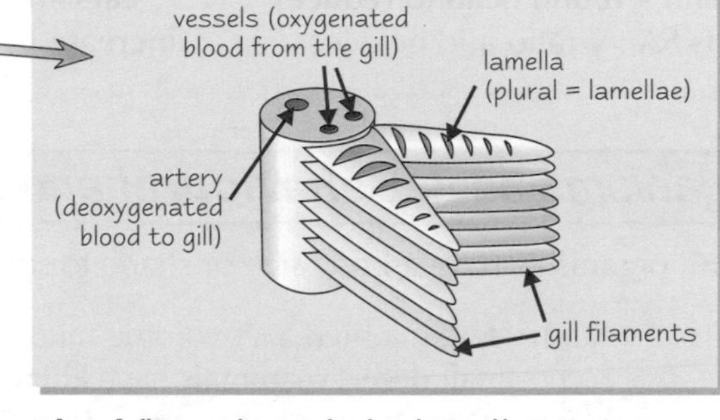


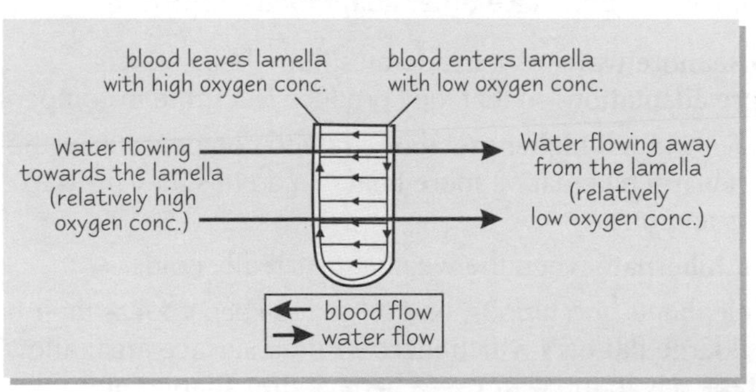


5) **Blood** flows through the lamellae in one direction and **water** flows over in the opposite direction. This is called a **counter-current system**. It maintains a **large concentration gradient** between the water and the blood — so as much oxygen as possible diffuses from the water into the blood.

## Insects use **Tracheae** to **Exchange Gases**

1) Insects have microscopic air-filled pipes called **tracheae** which they use for gas exchange.
2) Air moves into the tracheae through pores on the surface called **spiracles**.
3) **Oxygen** travels down the **concentration gradient** towards the **cells**. **Carbon dioxide** from the cells moves down its own concentration gradient towards the **spiracles** to be **released** into the atmosphere.
4) The tracheae branch off into smaller **tracheoles** which have **thin, permeable walls** and go to individual cells. This means that oxygen diffuses directly into the respiring cells — the insect's circulatory system doesn't transport $O_2$.
5) Insects use **rhythmic abdominal movements** to move air in and out of the spiracles.

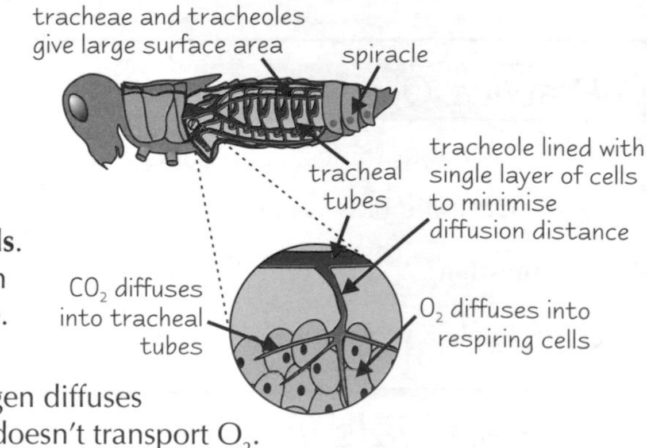

# Gas Exchange

## *Dicotyledonous Plants* Exchange Gases at the Surface of the *Mesophyll Cells*

1) Plants need $CO_2$ for **photosynthesis**, which produces $O_2$ as a waste gas. They need $O_2$ for **respiration**, which produces $CO_2$ as a waste gas.
2) The main gas exchange surface is the **surface of the mesophyll cells** in the leaf. They're well adapted for their function — they have a **large surface area**.
3) The mesophyll cells are inside the leaf. Gases move in and out through special pores in the **epidermis** called **stomata** (singular = stoma).
4) The stomata can **open** to allow exchange of gases, and **close** if the plant is losing too much water. **Guard cells** control the opening and closing of stomata.

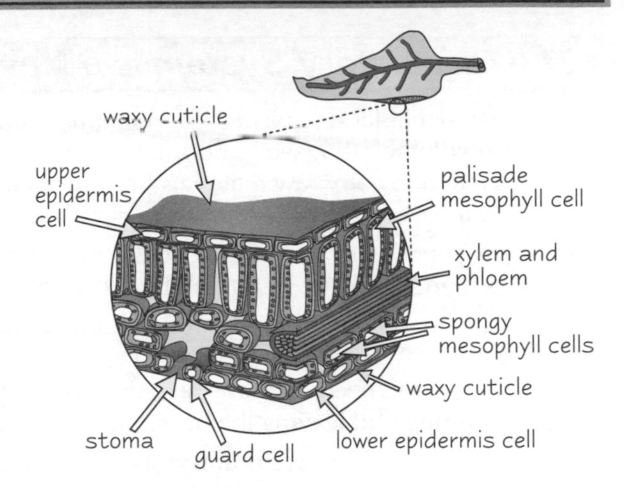


## *Insects* and *Plants* can *Control Water Loss*

Exchanging gases tends to make you **lose water** — there's a sort of **trade-off** between the two. Luckily for plants and insects though, they've evolved **adaptations** to **minimise water loss** without reducing gas exchange too much.

1) If **insects** are losing too much water, they **close** their **spiracles** using muscles. They also have a **waterproof, waxy cuticle** all over their body and **tiny hairs** around their spiracles, both of which **reduce evaporation**.
2) Plants' stomata are usually kept **open** during the day to allow **gaseous exchange**. Water enters the guard cells, making them **turgid**, which **opens** the stomatal pore. If the plant starts to get **dehydrated**, the guard cells lose water and become **flaccid**, which **closes** the pore.
3) Some plants are specially adapted for life in **warm**, **dry** or **windy** habitats, where **water loss** is a problem. These plants are called **xerophytes**.

See p. 75 for more on water loss in plants.

Examples of xerophytic adaptations include:

- Stomata sunk in **pits** which trap moist air, reducing evaporation.
- **Curled** leaves with the stomata inside, protecting them from wind.
- A layer of '**hairs**' on the epidermis to trap moist air round the stomata, reducing the concentration gradient of water.
- A **reduced number of stomata**, so there are fewer places for water to escape.
- **Waxy, waterproof cuticles** on leaves and stems to reduce evaporation.

sunken stomata
upper epidermis with thick waxy cuticle
lower epidermis
epidermal hairs trap humid air
curled leaf traps humid air and lowers exposed surface area

## *Practice Questions*

Q1 How are single-celled organisms adapted for efficient gas exchange?

Q2 What is the advantage to fish of having a counter-current system in their gills?

Q3 What are an insect's spiracles?

Q4 Through which pores are gases exchanged in plants?

**Exam Questions**

Q1 Give three ways gas exchange organs are adapted to their function. Give a different example for each one. [6 marks]

Q2 Explain why plants that live in the desert often have sunken stomata or stomata surrounded by hairs. [2 marks]

## *Keep revising and you'll be on the right trachea...*

*There's a pretty strong theme on these pages — whatever organism it is, to exchange gases efficiently it needs exchange organs with a large surface area, a thin exchange surface and a high concentration gradient. Don't forget that (or I'll hit you with a big stick).*

# The Circulatory System

*As the name suggests, the circulatory system is responsible for circulating stuff around the body— blood, to be specific. Most multicellular organisms (mammals, insects, fish, even French people) have a circulatory system of some type.*

## The **Circulatory System** is a **Mass Transport System**

1) Multicellular organisms, like **mammals**, have a **low surface area to volume ratio** (see p. 68), so they need a specialised **transport system** to carry raw materials from specialised **exchange organs** to their body cells — this is the **circulatory system**.
2) As you already know, the circulatory system is made up of the **heart** and **blood vessels**.
3) The heart **pumps blood** through blood vessels (arteries, arterioles, veins and capillaries) to reach different parts of the body. You need to **know** the names of **all** the blood vessels **entering** and **leaving** the **heart**, **liver** and **kidneys**.
4) Blood transports **respiratory gases**, products of **digestion**, **metabolic wastes** and **hormones** round the body.

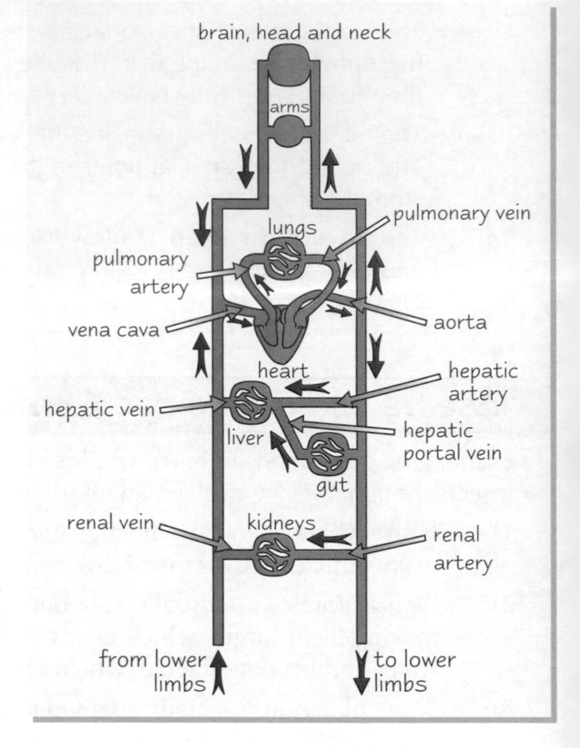


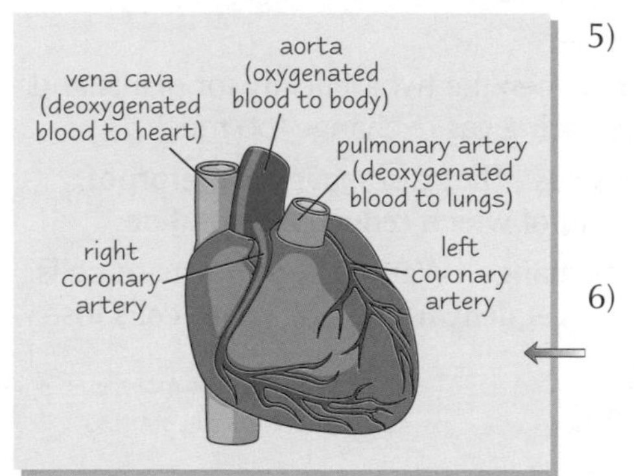


5) There are **two circuits**. One circuit takes blood from the **heart** to the **lungs**, then **back to the heart**. The other loop takes blood around the **rest of the body**.
6) The heart has its own blood supply — the left and right **coronary arteries**.

## **Different Blood Vessels** are Adapted for **Different Functions**

**Arteries**, **arterioles** and **veins** have different **characteristics**, and you need to know **why**...

1) **Arteries** carry blood **from** the heart **to** the rest of the body. Their walls are thick and **muscular** and have elastic tissue to cope with the **high pressure** produced by the heartbeat. The inner lining (**endothelium**) is **folded**, allowing the artery to **expand** — this also helps it to cope with high pressure. All arteries carry **oxygenated** blood except for the **pulmonary arteries**, which take deoxygenated blood to the lungs.

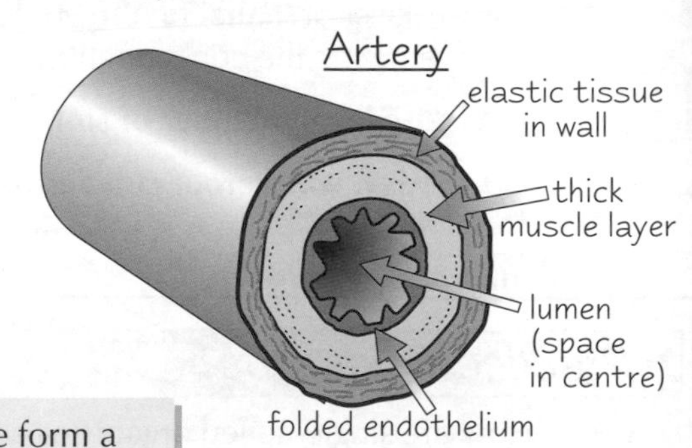


2) Arteries divide into smaller vessels called **arterioles**. These form a network throughout the body. Blood is directed to different **areas of demand** in the body by **muscles** inside the arterioles, which contract to restrict the blood flow or relax to allow full blood flow.

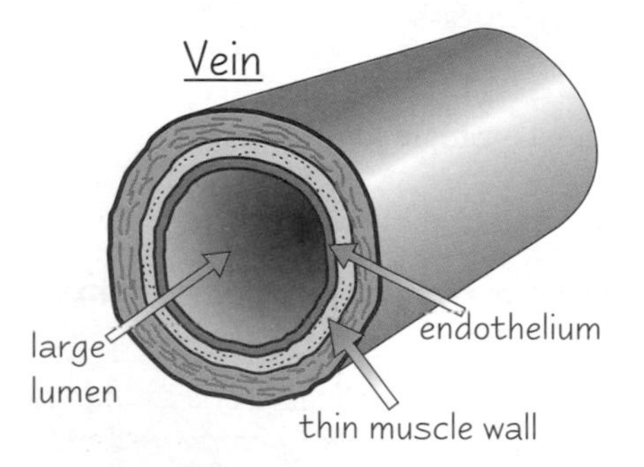


3) **Veins** take blood back **to the heart** under **low pressure**. They have a **wider** lumen than equivalent arteries, with very little elastic or muscle tissue. Veins contain **valves** to stop the blood flowing backwards. Blood flow through the veins is helped by contraction of the **body muscles** surrounding them. All veins carry **deoxygenated** blood (because oxygen has been used up by body cells), except for the **pulmonary veins**, which carry oxygenated blood to the heart from the lungs.

# The Circulatory System

## *Substances are Exchanged between Blood and Body Tissues at Capillaries*

Arterioles branch into **capillaries**, which are the **smallest** of the blood vessels. Substances (e.g. glucose and oxygen) are **exchanged** between cells and capillaries, so they're adapted for **efficient diffusion**.

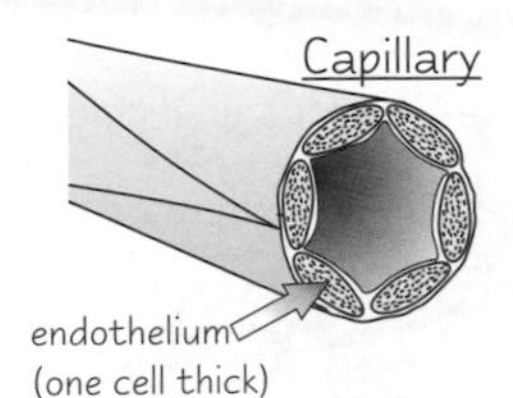


1) They're always found very **near cells in exchange tissues** (e.g. alveoli in the lungs), so there's a **short diffusion pathway**.
2) Their walls are only **one cell thick**, which also shortens the diffusion pathway.
3) There are a large number of capillaries, to **increase surface area** for exchange. Networks of capillaries in tissue are called **capillary beds**.

## *Tissue Fluid is Formed from Blood*

**Tissue fluid** is the fluid that **surrounds cells** in tissues. It's made from substances that leave the blood, e.g. oxygen, water and nutrients. Cells take in oxygen and nutrients from the tissue fluid, and release metabolic waste into it. Substances move out of blood capillaries, into the tissue fluid, by **pressure filtration**:

1) At the **start** of the capillary bed, nearest the arteries, the pressure inside the capillaries is **greater** than the pressure in the tissue fluid. This difference in pressure **forces fluid out** of the **capillaries** and into the **spaces** around the cells, forming tissue fluid.
2) As fluid leaves, the pressure reduces in the capillaries — so the pressure is much **lower** at the **end** of the capillary bed that's nearest to the veins.
3) Due to the fluid loss, the **water potential** at the end of the capillaries nearest the veins is **lower** than the water potential in the **tissue fluid** — so some **water re-enters** the capillaries from the tissue fluid at the vein end by **osmosis** (see p. 28 for more on osmosis).

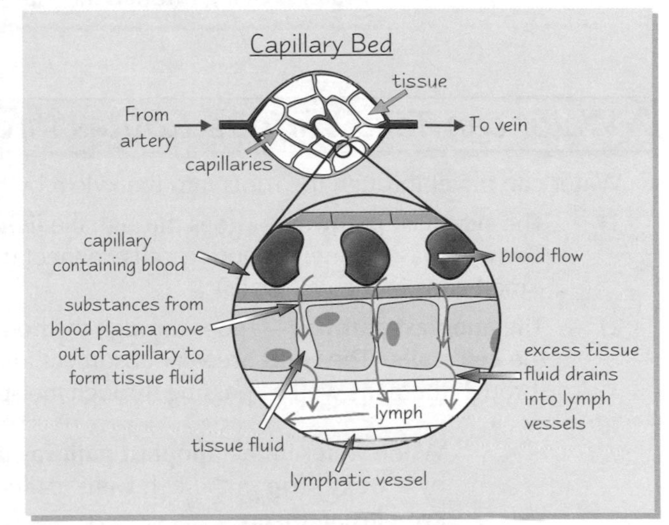


Unlike blood, tissue fluid **doesn't** contain **red blood cells** or **big proteins**, because they're **too large** to be pushed out through the capillary walls. Any **excess** tissue fluid is drained into the **lymphatic system** (a network of tubes that acts a bit like a drain), which transports this excess fluid from the tissues and dumps it back into the circulatory system.

## *Practice Questions*

Q1 Name all the blood vessels entering and leaving the heart.

Q2 Name all the blood vessels entering and leaving the kidney.

Q3 List four types of blood vessel.

Q4 Explain why capillaries are important in metabolic exchange.

**Exam Questions**

Q1 Describe the structure of an artery and explain how it relates to its function. [6 marks]

Q2 Explain how tissue fluid is formed and how it is returned to the circulation. [4 marks]

## *If blood can handle transport this efficiently, the trains have no excuse...*

*Four hours I was waiting at the train station this weekend. Four hours! Anyway, you may have noticed that biologists are obsessed with the relationship between structure and function, so whenever you're learning the structure of something, make sure you know how this relates to its function. Like veins, arteries and capillaries on these pages, for example.*

# Water Transport in Plants

*Water enters a plant through its roots and eventually, if it's not used, exits via the leaves. "Ah-ha," I hear you cry, "but how does it flow upwards, against gravity?" Well that, my friend, is a mystery that's about to be explained.*

## ***Water*** *Enters a Plant through its* ***Root Hair Cells***

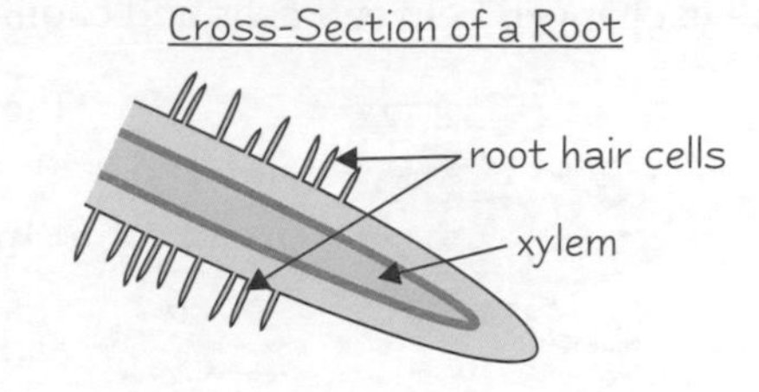


1) Water has to get from the **soil**, through the **root** and into the **xylem** — the system of vessels that **transports** water throughout the plant.
2) The bit of the root that absorbs water is covered in **root hairs**. These **increase** the root's **surface area**, speeding up water uptake.
3) Once it's absorbed, the water has to get through the **cortex**, including the **endodermis**, before it can reach the xylem.

Water always moves from areas of **higher water potential** to areas of **lower water potential** — it goes down a **water potential gradient**. The **soil** around roots generally has a **high water potential** (i.e. there's lots of water there) and **leaves** have a **lower water potential** (because water constantly **evaporates** from them). This creates a water potential gradient that keeps water moving through the plant in the right direction, **from roots to leaves**.

## *Water can Take* ***Various Routes*** *through the* ***Root***

The prison had been strangely quiet ever since plasmodesmata were installed.

Water can travel through the roots into the xylem by two different paths:

1) The **symplast pathway** — goes through the **living** parts of cells — the **cytoplasm**. The cytoplasm of neighbouring cells connect through **plasmodesmata** (small gaps in the cell walls).
2) The **apoplast pathway** — goes through the **non-living** parts of the root — the **cell walls**. The walls are very absorbent and water can simply **diffuse** through them, as well as passing through the spaces between them.

- When water in the **apoplast pathway** gets to the **endodermis** cells though, its path is blocked by a **waxy strip** in the cell walls, called the **Casparian strip**. Now the water has to take the **symplast pathway**.
- This is useful, because it means the water has to go through a **cell membrane**. Cell membranes are able to control whether or not substances in the water get through (see p. 26).
- Once past this barrier, the water moves into the **xylem**.

3) Both pathways are used, but the main one is the **apoplast pathway** because it provides the **least resistance**.

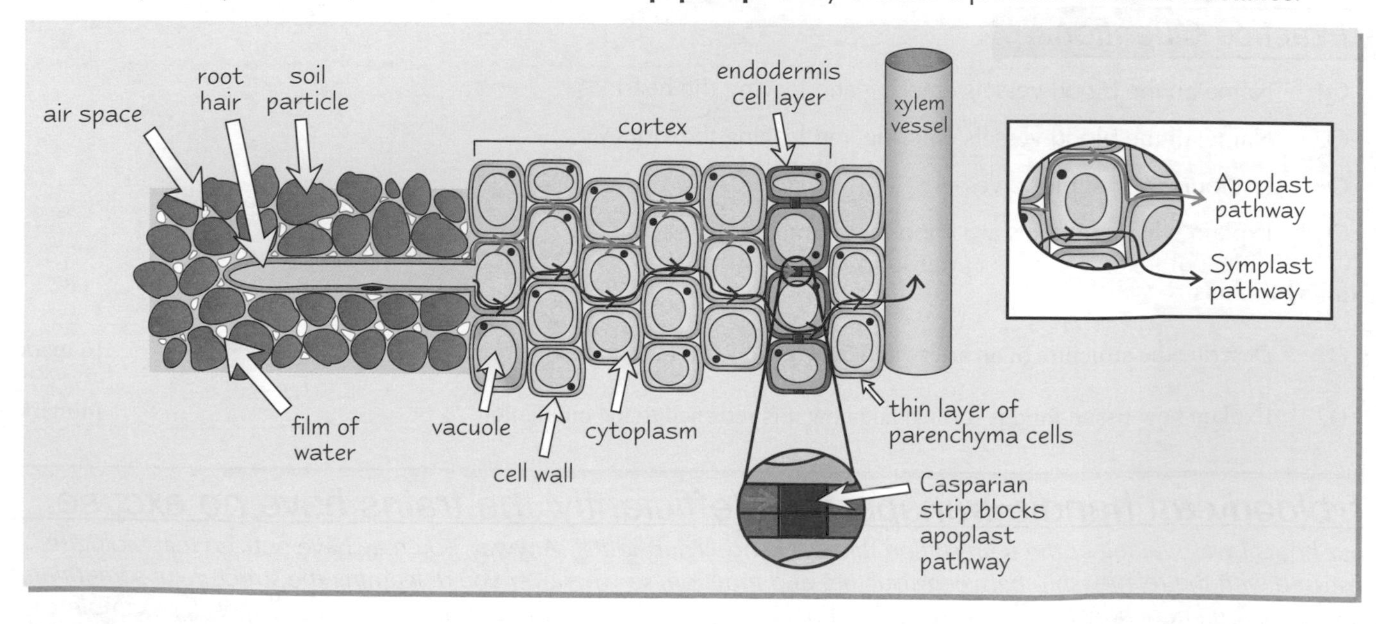

# Water Transport in Plants

## Water Moves Up a Plant Against the Force of Gravity

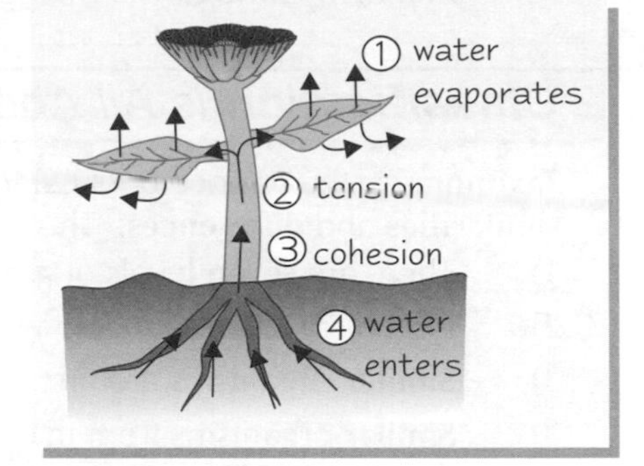


Water can move up a plant in two ways:

① **Cohesion** and **tension** help water move up plants, from roots to leaves, against the force of gravity.

1) Water **evaporates** from the **leaves** at the 'top' of the xylem. →
2) This creates **tension** (**suction**), which pulls more water into the leaf.
3) Water molecules are **cohesive** (they **stick together**) so when some are pulled into the leaf others follow. This means the whole **column** of water in the **xylem**, from the leaves down to the roots, **moves upwards**.
4) **Water** enters the stem through the **roots**.

② **Root pressure** also helps move the water upwards. When water is transported into the xylem in the roots, it creates a **pressure** and **shoves** water already in the xylem **further upwards**. This pressure is **weak**, and couldn't move water to the top of bigger plants by itself. But it helps, especially in young, small plants where the leaves are still developing.

## Transpiration is Loss of Water from a Plant's Surface

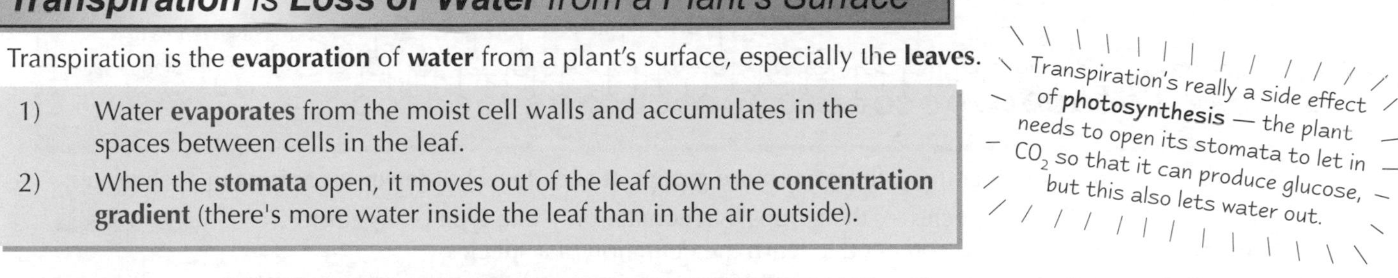

Transpiration is the **evaporation** of **water** from a plant's surface, especially the **leaves**.

1) Water **evaporates** from the moist cell walls and accumulates in the spaces between cells in the leaf.
2) When the **stomata** open, it moves out of the leaf down the **concentration gradient** (there's more water inside the leaf than in the air outside).

Transpiration's really a side effect of **photosynthesis** — the plant needs to open its stomata to let in $CO_2$ so that it can produce glucose, but this also lets water out.

## Four Main Factors Affect Transpiration Rate

1) **Light** — the **lighter** it is the **faster** the **transpiration rate**. This is because the **stomata open** when it gets **light**. When it's **dark** the stomata are usually **closed**, so there's little transpiration.
2) **Temperature** — the **higher the temperature** the **faster** the **transpiration rate**. Warmer water molecules have more energy so they **evaporate** from the cells inside the leaf **faster**. This **increases** the **concentration gradient** between the inside and outside of the leaf, making water **diffuse out** of the leaf **faster**.
3) **Humidity** — the **lower** the **humidity**, the **faster** the **transpiration rate**. If the air around the plant is **dry**, the **concentration gradient** between the leaf and the air is **increased**, which increases transpiration.
4) **Wind** — the **windier** it is, the **faster** the **transpiration rate**. Lots of air movement **blows away** water molecules from around the stomata. This **increases** the **concentration gradient**, which increases the rate of transpiration.

## Practice Questions

Q1 How does water enter a plant?

Q2 Explain how root pressure helps to move water through plants.

Q3 Give four factors that affect transpiration rate.

**Exam Questions**

Q1 Describe the two routes water can take through the roots of a plant. [4 marks]

Q2 Explain why movement of water in the xylem stops if the leaves of a plant are removed. [4 marks]

## So many routes through the roots...

*Lots of impressive biological words on this page to amaze your friends and confound your enemies. Go through all this stuff again, and whenever you come across a ridiculous word like 'plasmodesmata', just stop and check you know exactly what it means. (Personally, I think they should've just called them 'cell wall gaps', but nobody ever listens to me.)*

# Principles of Classification

*For hundreds of years people have been putting organisms into groups to make it easier to recognise and name them. For example, my brother is a member of the species Idioto bigearian (Latin for idiots with big ears).*

## *Classification* is All About *Grouping Together Similar Organisms*

**Taxonomy** is the science of classification. It involves **naming** organisms and **organising them** into **groups** based on their **similarities** and **differences**. This makes it **easier** for scientists to **identify** and **study** them.

1) There are **seven** levels of groups (called taxonomic groups) used to classify organisms.
2) Organisms can only belong to **one group** at **each level** in the taxonomic hierarchy — there's **no overlap**.
3) **Similar organisms** are first sorted into **large groups** called **kingdoms**, e.g. all animals are in the animal kingdom.
4) **Similar** organisms from that kingdom are then grouped into a **phylum**. **Similar** organisms from each phylum are then grouped into a **class**, and **so on** down the seven levels of the hierarchy.

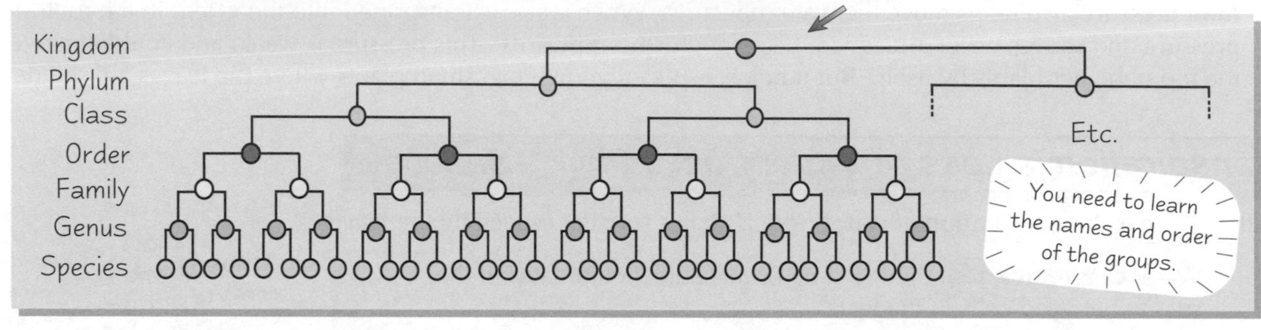


5) As you move **down** the hierarchy, there are **more groups** at each level but **fewer organisms** in each group.
6) The hierarchy **ends** with **species** — the groups that contain only **one type** of organism (e.g. humans, dogs, *E. coli*). You need to **learn** the definition of a **species**:

**A species is a group of similar organisms able to reproduce to give fertile offspring.**

Species are given a **scientific name** to **distinguish** them from similar organisms. This is a **two-word** name in **Latin**. The **first** word is the **genus** name and the **second** word is the **species** name — e.g. humans are *Homo sapiens*. Giving organisms a scientific name enables scientists to **communicate** about organisms in a standard way that minimises confusion. E.g. Americans call a type of bird **cockatoos** and Australians call them **flaming galahs** (best said with an Australian accent), but it's the **same bird**. If the correct **scientific name** is used — *Eolophus roseicapillus* — there's no confusion.

7) Scientists constantly **update** classification systems because of **discoveries** about new species and new **evidence** about known organisms (e.g. **DNA sequence** data — see p. 78).

## *Phylogenetics* Tells Us About an Organism's *Evolutionary History*

1) **Phylogenetics** is the study of the **evolutionary history** of groups of **organisms**.
2) All organisms have **evolved** from shared common ancestors (**relatives**). E.g. members of the Hominidae family (great apes and humans) evolved from a common ancestor. First orangutans **diverged** (evolved to become a **different species**) from this common ancestor. Next gorillas diverged, then humans, closely followed by bonobos and chimpanzees.
3) Phylogenetics tells us **who's related** to whom and how **closely related** they are.
4) Closely related species **diverged** away from each other **most recently**. E.g. the phylogenetic tree opposite shows the **Hominidae tree**. Humans and **chimpanzees** are **closely** related, as they diverged very **recently**. You can see this because their branches are **close** together. Humans and orangutans are more **distantly** related, as they diverged longer ago, so their branches are **further** apart.

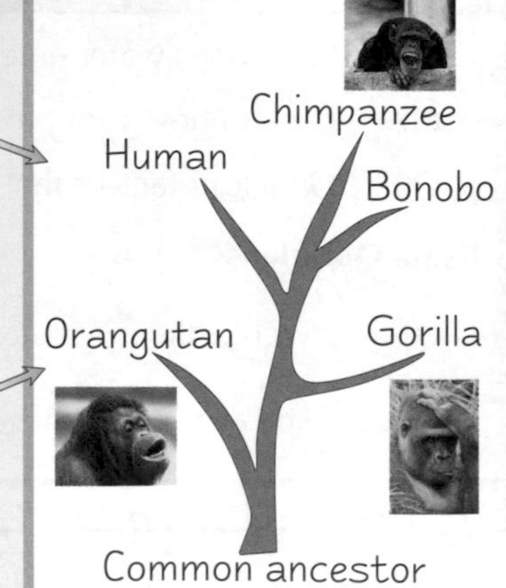

# Principles of Classification

## **Defining** Organisms as **Distinct Species** Can be Quite **Tricky**

1) Scientists can have problems when using the **definition** of a species on the previous page to decide which species an organism belongs to or if it's a new, **distinct species.**
2) This is because you can't always see their **reproductive behaviour** (you can't always tell if different organisms can reproduce to give **fertile offspring**).
3) Here are some of the **reasons** why you can't always see their reproductive behaviour:

> 1) They're **extinct**, so obviously you **can't** study their reproductive behaviour.
> 2) They **reproduce asexually** — they never **reproduce together** even if they belong to the same species, e.g. bacteria.
> 3) There are **practical** and **ethical issues** involved — you can't see if some organisms reproduce successfully in the wild (due to geography) and you can't study them in a lab (because it's unethical), e.g. humans and chimps are classed as separate species but has anyone ever tried mating them?

Evidence has been found of human/parrot reproduction.

4) Because of these problems some organisms are **classified** as one species or another using other **techniques**.
5) Scientists can now compare the **DNA** (see p. 78) of organisms to see **how related** they are, e.g. the **more** DNA they have in common the more **closely related** they are. But there's no strict cut-off to say **how much** shared DNA can be used to define a **species**. For example, **only** about 6% of human DNA **differs** from chimpanzee DNA but we are separate species.

## *Practice Questions*

Q1 What is taxonomy?

Q2 Why is taxonomy and the classification system important?

Q3 What name is given to the groups in the taxonomic hierarchy that contain the largest number of species?

Q4 Why is the scientific naming system important?

Q5 What is phylogenetics?

Deer, Sheep, Buffalo, Llama, Camel, Pig, Human, Rhesus Monkey, Green Monkey, Rabbit, Dog, Cat

**Exam Questions**

Q1 The phylogenetic tree on the right shows the evolutionary history of some mammalian species.

a) Which species shown on the tree is most closely related to humans? Explain how you know this. [2 marks]

b) Which animal is a close relative of both camels and deer? [1 mark]

Q2 Define a species. [2 marks]

Q3 Complete the table for the classification of humans. [5 marks]

| | Phylum | | | Family | | |
|---|---|---|---|---|---|---|
| Animalia | Chordata | Mammalia | Primates | Hominidae | Homo | sapiens |

Q4 Give three reasons why it can be hard for scientists to define organisms as members of a distinct species. [3 marks]

## *Phylum — I thought that was the snot you get with a cold...*

*If you don't understand what a species is, you'll find yourself struggling on the next few pages. So, all together now... a species is a group of similar organisms able to reproduce to give fertile offspring... Now, don't forget that.*

# Classifying Species

*Early classification systems only used observable features to place organisms into groups, e.g. six heads, four toes, a love of Take That. But now a variety of evidence is used to classify organisms...*

## *Species* Can be *Classified* by *Their DNA* or *Proteins*

1) Species can be **classified** into different groups in the **taxonomic hierarchy** (see p. 76) based on **similarities** and **differences** in their **genes**.
2) This can be done by comparing their **DNA sequence** or by looking at their **proteins** (which are coded for by their DNA).
3) Organisms that are **more closely** related will have **more similar** DNA and proteins than distantly related organisms.

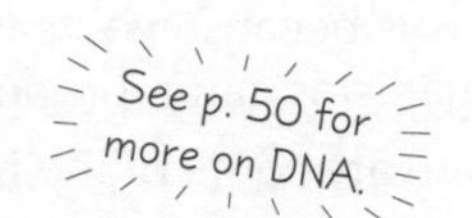


## *DNA* Can be Compared *Directly* or by Using *Hybridisation*

DNA similarity can be measured by looking at the **sequence of bases** or by **DNA hybridisation**:

**DNA sequencing**

The **DNA** of organisms can be directly compared by looking at the **order of the bases** (As, Ts, Gs and Cs) in each. Closely related species will have a **higher percentage** of similarity in their DNA **base order**, e.g. humans and chimps share around 94%, humans and mice share about 85%.
DNA sequence comparison has led to **new classification systems** for **plants**, e.g. the classification system for flowering plants is based almost entirely on **similarities** between DNA sequences.

**DNA Hybridisation**

DNA Hybridisation is used to see how similar DNA is **without** sequencing it. Here's how it's done:

1) DNA from **two** different species is collected, separated into **single strands** and **mixed** together.
2) Where the **base sequences** of the DNA are the same on both strands, **hydrogen bonds** form between the base pairs by **specific base pairing**.
The more DNA bases that **hybridise** (bond) together, the more **alike** the DNA is.
3) The DNA is then **heated** to separate the strands again. **Similar DNA** will have **more hydrogen bonds** holding the two strands together so a **higher temperature** (i.e. **more energy**) will be needed to separate the strands.

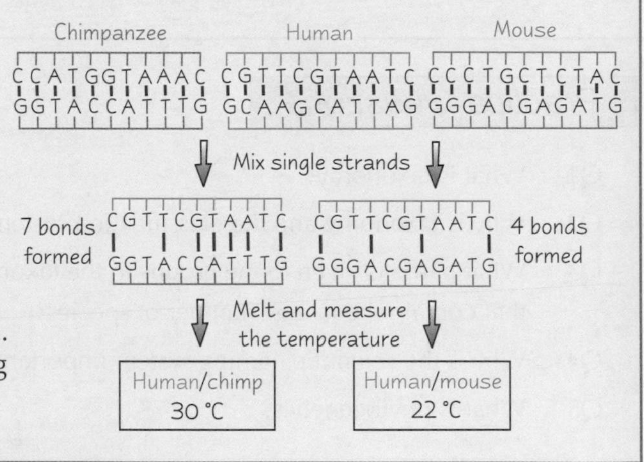


## *Proteins* Can be Compared *Directly* or by Using *Immunology*

Similar organisms will have **similar proteins** in their cells. Proteins can be compared in **two** ways:

1) **Comparing amino acid sequence**
Proteins are made of **amino acids**. The **sequence** of amino acids in a protein is coded for by the **base sequence** in DNA (see p. 52). **Related organisms** have similar DNA sequences and so **similar amino acid sequences** in their proteins.

2) **Immunological comparisons** — Similar proteins will bind the same **antibodies** (see p. 6). E.g. if antibodies to a **human version** of a protein are added to isolated samples from some other **species**, any protein that's like the human version will also be **recognised** (bound) by that antibody.

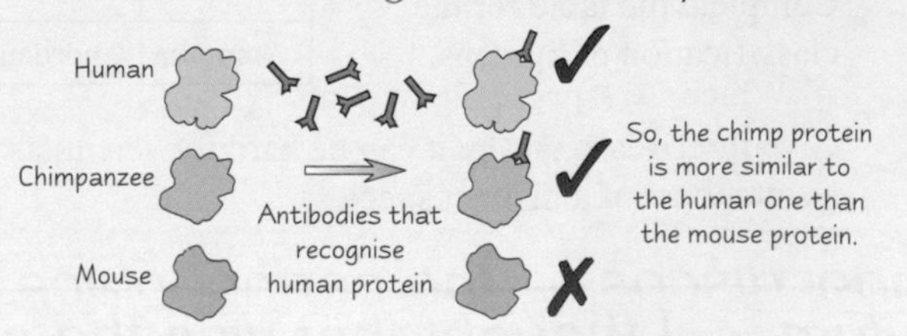

# Classifying Species

## You Need to be Able to *Interpret Data* on DNA and Protein *Similarities*

Here are two examples of the kind of thing you might get:

| | Species A | Species B | Species C | Species D |
|---|---|---|---|---|
| Species A | 100% | 86% | 42% | 44% |
| Species B | 86% | 100% | 51% | 53% |
| Species C | 42% | 51% | 100% | 91% |
| Species D | 44% | 53% | 91% | 100% |

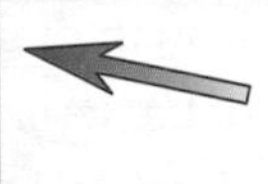

The table on the left shows the **% similarity of DNA** using DNA sequence analysis between several species of bacteria. The data shows that species **A** and **B** are **more closely related** to each other than they are to either C or D. Species **C** and **D** are also **more closely related** to each other than they are to either A or B.

The diagram on the right shows the **amino acid sequences** of a certain protein from three different species.

You can see that the amino acid sequences from species **A** and **B** are **very similar**. The sequence from species **C** is **very different** to any of the other sequences. This would suggest that species **A** and **B** are **more closely related**.

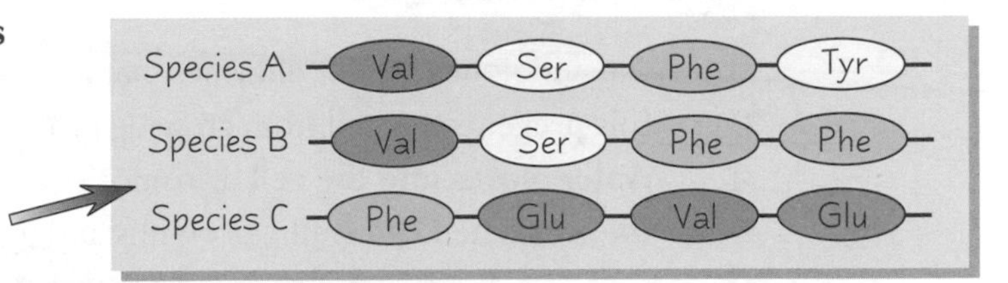


## *Courtship Behaviour* can Also be Used to *Classify Species*

1) **Courtship behaviour** is carried out by organisms to **attract** a mate of the **right species**.
2) It can be fairly simple, e.g. **releasing chemicals**, or quite complex, e.g. a series of **displays**.
3) Courtship behaviour is **species specific** — only members of the same species will do and respond to that courtship behaviour. This prevents **interbreeding** and so makes reproduction **more successful** (as mating with the wrong species won't produce **fertile** offspring).
4) Because of this specificity, courtship behaviour can be used to **classify organisms**.
5) The more **closely related** species are, the **more similar** their courtship behaviour.

Some examples of courtship behaviour include:

1) **Fireflies** give off **pulses of light**. The pattern of flashes is specific to each species.
2) **Crickets** make **sounds** that are similar to Morse code, the code being different for different species.
3) **Male peacocks** show off their **colourful tails**. This tail pattern is only found in peacocks.
4) **Male butterflies** use **chemicals** to attract females. Only those of the correct species respond.

Geoff's jive never failed to attract a mate.

## Practice Questions

Q1 Suggest two ways that DNA from two different species could be compared.

Q2 Suggest two ways that proteins from two different species could be compared.

**Exam Questions**

Q1 Explain how DNA hybridisation is used to analyse similarities between the DNA of two species. [5 marks]

Q2 The amino acid sequence of a specific protein was used to make comparisons between four species of animal. The results are shown on the right.

| Species | Amino acid 1 | Amino acid 2 | Amino acid 3 | Amino acid 4 |
|---|---|---|---|---|
| Rabbit | His | Ala | Asp | Lys |
| Mouse | Thr | Ala | Asp | Val |
| Chicken | Ala | Thr | Arg | Arg |
| Rat | Thr | Ala | Asp | Phy |

a) Which two species are the most closely related? [1 mark]

b) Which species is the most distantly related to the other three? [1 mark]

## School discos — the perfect place to observe courtship behaviour...

*It's important that you understand that the more similar the DNA and proteins, the more closely related (and hence the more recently diverged) two species are. This is because relatives have similar DNA, which codes for similar proteins, made of a similar sequence of amino acids. Just like you and your family — you're all alike because your DNA's similar.*

# Antibiotic Action and Resistance

*These pages are all about antibiotics and how they kill (or inhibit) bacteria. But don't feel sorry for the bacteria — they're getting their own back by evolving antibiotic resistance. Sneaky...*

## ***Antibiotics*** *Are Used to* ***Treat Bacterial Diseases***

1) Antibiotics are **chemicals** that either **kill** or **inhibit** the **growth** of bacteria.
2) **Different types** of antibiotics kill or inhibit the growth of bacteria in **different ways**.
3) Some **prevent growing** bacterial cells from **forming** the bacterial **cell wall**, which usually gives the cell structure and support (see p. 32).
4) This can lead to **osmotic lysis**:

> 1) The antibiotics **inhibit enzymes** that are needed to make the **chemical bonds** in the cell wall.
> 2) This **prevents** the cell from growing properly and **weakens** the cell wall.
> 3) **Water** moves **into the cell** by **osmosis**.
> 4) The **weakened cell wall** can't withstand the increase in **pressure** and **bursts** (**lyses**).

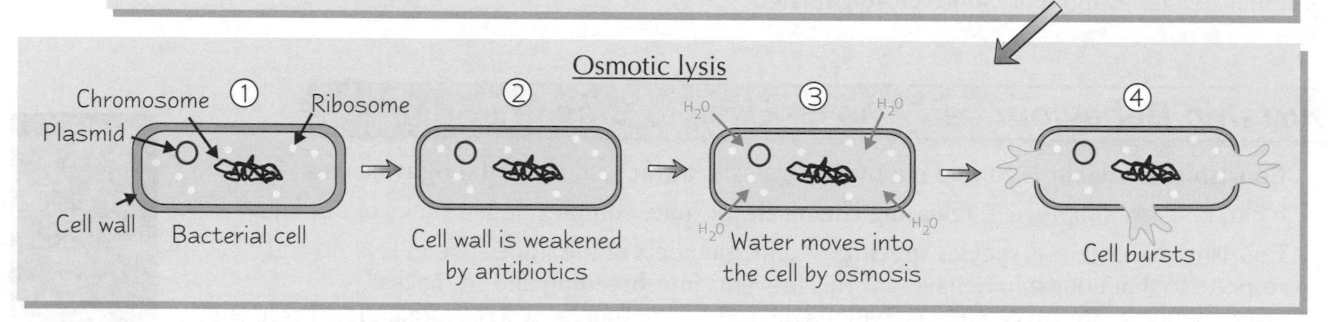


## ***Mutations*** *in* ***Bacterial DNA*** *can Cause* ***Antibiotic Resistance***

See p. 53 for more on DNA and mutations.

1) The **genetic material** in bacteria is the same as in most other organisms — **DNA**.
2) The DNA of an organism contains **genes** that carry the instructions for different **proteins**. These proteins determine the organism's **characteristics**.
3) **Mutations** are **changes** in the **base sequence** of an organism's DNA.
4) If a mutation occurs in the DNA of a gene it could change the protein and cause a **different characteristic**.
5) Some mutations in bacterial DNA mean that the bacteria are **not affected** by a particular antibiotic any more — they've developed **antibiotic resistance**.

**Example**

> **Methicillin** is an antibiotic that inhibits an enzyme involved in **cell wall formation** (see above). Some bacteria have developed resistance to methicillin, e.g. methicillin-resistant *Staphylococcus aureus* (**MRSA**). Usually, **resistance** to methicillin occurs because the **gene** for the **target enzyme** of methicillin has **mutated**. The mutated gene produces an **altered enzyme** that methicillin no longer **recognises**, and so **can't inhibit**.

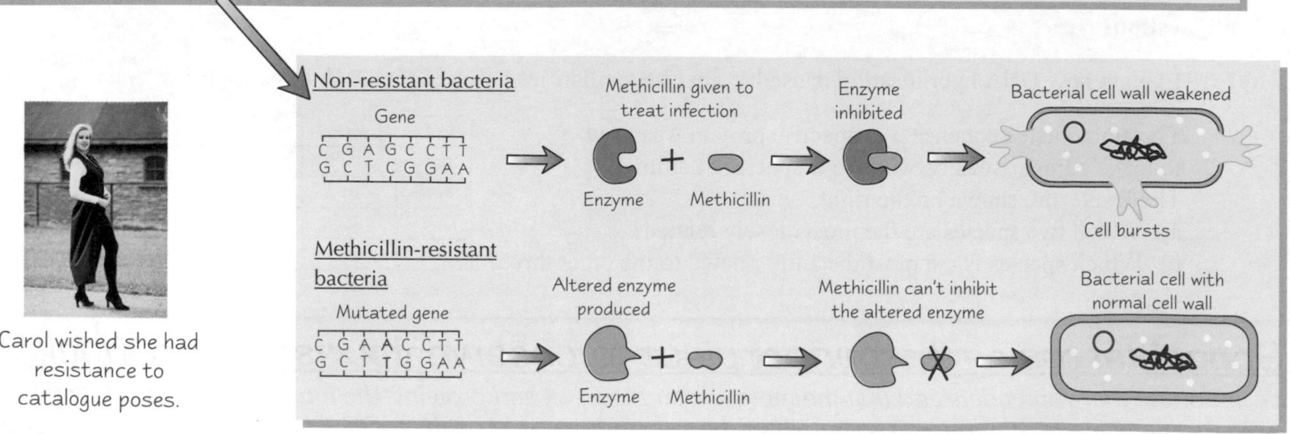


Carol wished she had resistance to catalogue poses.

# Antibiotic Action and Resistance

## *Antibiotic Resistance can be Passed On Vertically...*

**Vertical gene transmission** is where genes are passed on during **reproduction**.

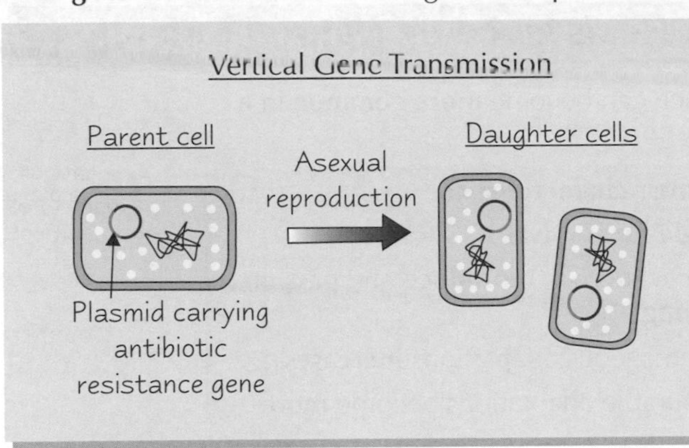


1) Bacteria reproduce **asexually**, so each daughter cell is an **exact copy** of the parent.
2) This means that **each** daughter cell has an exact copy of the parent cell's **genes**, including any that give it **antibiotic resistance**.
3) Genes for antibiotic resistance can be found in the bacterial **chromosome** or in **plasmids** (small **rings of DNA** found in bacterial cells, see p. 32).
4) The chromosome and any plasmids are passed on to the daughter cells during reproduction.

## *...or Horizontally*

1) Genes for resistance can also be passed on **horizontally**.
2) Two bacteria **join together** in a process called **conjugation** and a **copy of a plasmid** is passed from one cell to the other.
3) Plasmids can be passed on to a member of the **same species** or a totally **different species**.

Horizontal Gene Transmission

Bacterial cell with antibiotic resistance gene in plasmid

Bacteria without plasmid

Bacterial conjugation — copy of plasmid transferred through stalk called a pilus

Both bacterial cells have a copy of the plasmid containing the antibiotic resistance gene

## *Practice Questions*

Q1 What is osmotic lysis?

Q2 What is the genetic material in bacteria?

Q3 What is the difference between vertical and horizontal gene transmission?

**Exam Question**

Q1 More and more bacteria are becoming resistant to antibiotics such as penicillin.

a) Describe how resistance to an antibiotic arises in bacteria. [3 marks]

b) Describe how resistance to antibiotics is spread between two bacteria. [3 marks]

c) Penicillin is a cell wall inhibitor antibiotic. Explain how penicillin kills bacteria. [4 marks]

## *Horizontal gene transmission — that's not what it was called in my day...*

*There are lots of nice, colourful pictures on this page, but they're not just here to make the place look pretty you know. They're here to help you learn the different processes you need to understand for your exam — osmotic lysis, vertical transmission and horizontal transmission — so get scribblin' and learnin'. Go on, go on, go on...*

# Antibiotic Resistance

*Mutations arise by accident but if they're useful, e.g. give antibiotic resistance, then natural selection will make sure they're passed on and on and on and on and on and on and on (and on and on and on)...*

## Bacterial Populations Evolve Antibiotic Resistance by Natural Selection

An **adaptation** (a useful characteristic) like antibiotic resistance can become **more common** in a **population** because of **natural selection**:

1) Individuals within a population **show variation** in their **characteristics**.
2) **Predation**, **disease** and **competition** create a **struggle for survival**.
3) Individuals with **better adaptations** are **more likely** to **survive**, **reproduce** and **pass on** the **alleles** that cause the adaptations to their **offspring**.
4) Over time, the **number** of individuals with the advantageous adaptations **increases**.
5) Over generations this leads to **evolution** as the favourable adaptations become **more common** in the population.

Adaptations are caused by gene mutations.

Here's how populations of antibiotic-resistant bacteria evolve by natural selection:

1) Some individuals in a population have alleles that give them **resistance** to an **antibiotic**.
2) The population is **exposed** to that antibiotic, **killing** bacteria **without** the antibiotic resistance allele.
3) The **resistant bacteria survive** and **reproduce** without competition, passing on the allele that gives antibiotic resistance to their offspring.
4) After some time **most** organisms in the population will carry the antibiotic resistance allele.

## Natural Selection also Occurs in Other Organisms

1) Natural selection happens in **all populations** — not just in bacterial populations.
2) There are loads of examples, but they all follow the same basic principle — the organism has a **characteristic** that makes it more likely to **survive**, **reproduce** and pass on the **alleles** for the better characteristic.
3) In your **exam** you might be asked to explain why certain characteristics are common (or have increased).
4) To do this you should **identify** why the **adaptations** (**characteristics**) are useful and **explain** how they've become more common due to **natural selection**.
5) Here are some examples of the kinds of characteristics that can help organisms to survive:

| Adaptations that could increase chance of survival | How the adaptations could increase survival |
|---|---|
| Streamlined body, camouflage, larger paws for running quicker etc. | They help to escape from predators. |
| Streamlined body, camouflage, larger paws for running quicker, larger claws, longer neck etc. | They help to catch prey/get food. |
| Shorter/longer hairs, large ears, increased water storage capacity etc. | They make the animal more suited to the climate. |

Dave wasn't convinced his camouflage was working.

# Antibiotic Resistance

## Antibiotic Resistance Makes it Difficult to Treat Some Diseases

Diseases caused by bacteria are treated using **antibiotics**. Because bacteria are becoming resistant to different antibiotics through **natural selection** it's becoming more and more **difficult** to treat some bacterial infections, such as tuberculosis (**TB**) and methicillin-resistant *Staphylococcus aureus* (**MRSA**).

### Tuberculosis

1) **TB** is a **lung disease** caused by bacteria.
2) TB was once a **major killer** in the UK, but the number of people dying from TB **decreased** with the development of **specific antibiotics** that killed the bacterium. Also the number of people catching TB **dropped** due to a vaccine (see p. 8)
3) More recently, some populations of TB bacteria have **evolved** resistance to the **most effective** antibiotics. **Natural selection** has led to populations that are resistant to a **range** of different antibiotics — the populations (strains) are **multidrug-resistant**.
4) To try to combat the **emergence** of resistance, TB treatment now involves taking a **combination** of different antibiotics for about **6 months**.
5) TB is becoming harder to treat as multidrug-resistant strains are **evolving quicker** than **drug companies** can develop new antibiotics.

There's more about TB on page 36.

### MRSA

1) **Methicillin-resistant *Staphylococcus aureus*** (MRSA) is a strain of the *Staphylococcus aureus* bacterium that has evolved to be resistant to a number of commonly used antibiotics, including **methicillin.**
2) *Staphylococcus aureus* causes a **range** of illnesses from **minor skin infections** to **life-threatening diseases** such as **meningitis** and **septicaemia**.
3) The major problem with MRSA is that some strains are resistant to **nearly all** the antibiotics that are available.
4) Also, it can take a long time for **clinicians** to determine which antibiotics, if any, will **kill** the strain each individual is infected with. During this time the **patient** may become **very ill** and even **die**.
5) **Drug companies** are trying to **develop alternative ways** of treating MRSA to try to combat the emergence of resistance.

Lab tests are carried out to see if any antibiotics can kill a strain of MRSA.

## Practice Questions

Q1 Briefly describe the process of natural selection.

Q2 Why is it becoming more difficult to treat TB infections?

**Exam Questions**

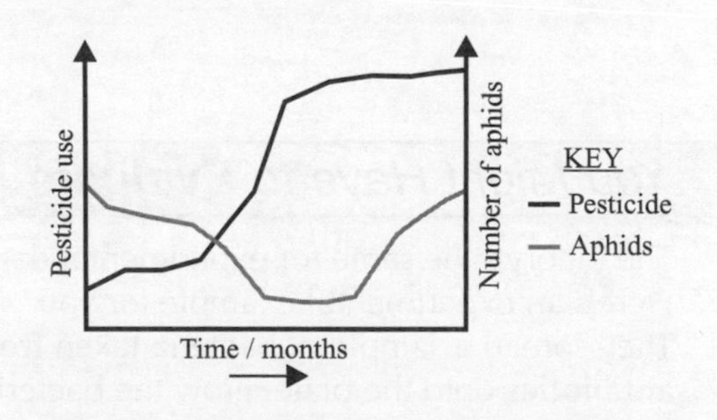


Q1 The graph shows the use of an anti-aphid pesticide on a farm and the number of aphids found on the farm over a period of time.

Describe and explain the change in aphid numbers shown in the graph. [6 marks]

Q2 The bat *Anoura fistulata* has a very long tongue (up to one and a half times the length of its body). The tongue enables the bat to feed on the nectar inside a deep tubular flower found in the forests of Ecuador.

Describe how natural selection can explain the evolution of such a long tongue. [3 marks]

## Why do giraffes have long necks?...*

*Adaptation and selection aren't that bad really... just remember that any characteristic that increases the chances of an organism getting more dinner, getting laid or avoiding being gobbled up by another creature will increase in the population (due to the process of natural selection). Now I know why mullets have disappeared... so unattractive...*

**So they can reach food found high up. This means they're more likely to survive, reproduce and pass on their alleles (genes). So no, it's not because they've got smelly feet.*

# Evaluating Resistance Data

*The number of infections caused by resistant bacteria is rising, so it's important to keep an eye on them and any new ones that pop up. Like they say, you need to know your enemies if you want to beat them (OK, so I don't know who says it, but someone does. I think I heard it on Catchphrase once...).*

## *You Need to be Able to* **Evaluate Data** *About* **Antibiotic Resistance**

It's very possible that you could get some data from a **study** into antibiotic resistance in the exam.
You need to be able to **evaluate** the **methodology**, **data** and any **conclusions** drawn.
Here's an example:

This study investigated the **number** of **death certificates mentioning** *Staphylococcus aureus (**S. aureus**)* and methicillin-resistant *S. aureus* (**MRSA**) in the UK between 1993 and 2002. The data was collected from **UK death certificates** issued between 1993 and 2002. The **results** are shown in the graph opposite.

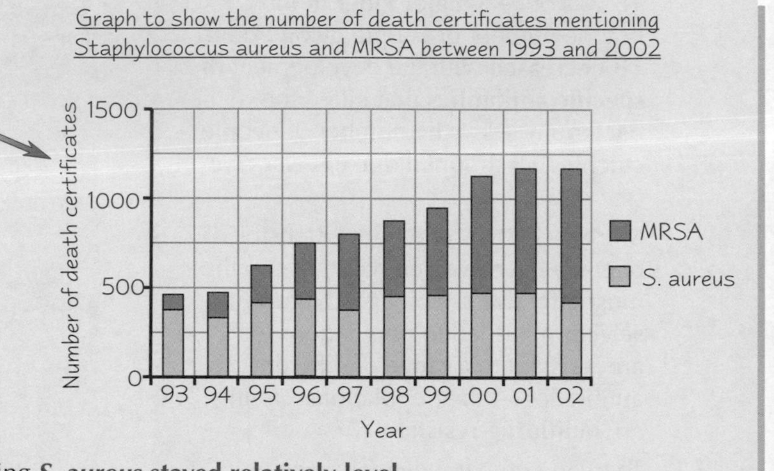


Here are some of the things you might be asked to do:

1) **Describe the data** — This study shows that the number of death certificates mentioning **all forms** of *S. aureus* **increased** between 1993 and 2002. The number mentioning **MRSA increased** while the number mentioning ***S. aureus*** **stayed relatively level**.
2) **Check the evidence backs up any conclusions** — Dr Bottril said, 'This data shows that the number of deaths **caused** by MRSA is rising'. Does the data support this conclusion? No. The study only looked at death certificates **mentioning** MRSA, not deaths **caused** by MRSA.
3) **Other points to consider**
   - This study looked at the number of **death certificates** mentioning MRSA. Studying the number of **reported** MRSA **infections** each year may have been a **better way** of investigating the occurrence of bacterial resistance.
   - Some death certificates may not have mentioned MRSA because it wasn't the **cause of death**, but the people may have been **infected** at the time. This means the data **doesn't** reflect the number of infections.
   - Increased **awareness** of MRSA may have influenced the decision to include MRSA on the death certificate, **biasing** the data.

## *You Might Have to* **Evaluate Experimental Data**

The theory's the same for experimental data — **evaluate** the **methodology**, **data** and **conclusions**.
Here's an experimental example for you — A clinician needs to find out which antibiotics will treat a **patient's infection**. They spread a sample of bacteria taken **from the patient** onto an agar plate. Then they place paper discs **soaked** with **antibiotics** onto the plate, grow the bacteria and **measure the areas of growth inhibition** after a set period of time:

1) **Draw conclusions** — Be **precise** about what the data shows. A 250 mg dose of **streptomycin inhibited** growth the most. A 250 mg dose of **tetracycline inhibited** growth a small amount. The bacteria appear to be **resistant** to **methicillin** up to **250 mg**.
2) **Evaluate the methodology** — the experiment included a **negative control**, which is good. The negative control is a paper disc soaked in sterile water. The bacteria grew around this disc, which shows the paper disc **alone** doesn't kill the bacteria.

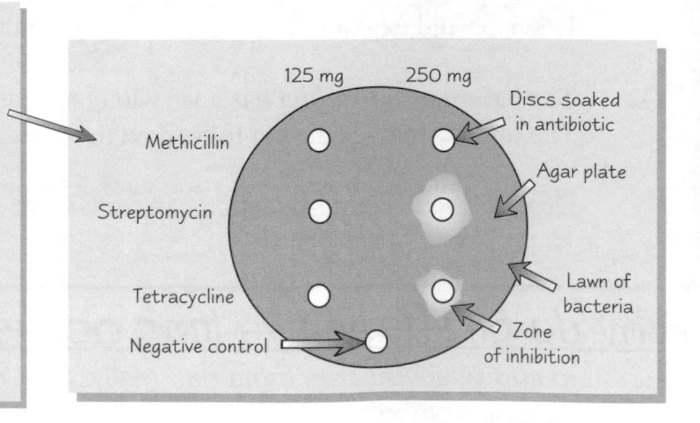

# Evaluating Resistance Data

## Decisions are Made Using Scientific Knowledge

Bacteria **will develop** antibiotic resistance by natural selection — it's nature. But scientific research has shown that certain things can be done to **slow down** the natural process. People working in the **public health sector**, together with patients, have to be made aware of recent **scientific findings** so that they can **act** upon them.
Here are two examples:

**Scientific Knowledge:** Using an **antiseptic gel** to wash hands can help to **reduce the spread** of infectious diseases by **person-to-person contact**.
**Decision:** Health workers should **reduce spread** by washing their hands with **antiseptic gel** (placed at all hand basins) before and after **visiting each patient** on a ward.

**Scientific knowledge:** Bacteria become resistant to antibiotics **more quickly** when antibiotics are **misused** and patients **don't finish the course**.
**Decision:** Doctors should only prescribe antibiotics when **absolutely necessary**. Patients have to be told the **importance** of finishing **all the antibiotics** even if they start to feel better.

## There are Ethical Issues Surrounding the Use of Antibiotics

People are very concerned about the spread of antibiotic-resistant bacteria. **Limiting the use** of antibiotics is one way of helping to slow down the emergence of resistance, but this raises some **ethical issues**.

1) Some people believe that antibiotics should only be used in **life-threatening situations** to reduce the increase of resistance. Others argue against this because people would take **more time off work** for illness, it could **reduce** people's **standard of living**, it could increase the **incidence of disease** and it could cause **unnecessary suffering**.
2) A few people believe doctors shouldn't prescribe antibiotics to those suffering **dementia**. They argue that they may forget to take them, increasing the chance of **resistance** developing. However, some people argue that **all patients** have the **right to medication**.
3) Some also argue that **terminally ill** patients shouldn't receive antibiotics because they're going to die. But **withholding** antibiotics from these patients could reduce their **length of survival** and **quality of life**.
4) Some people believe **animals shouldn't** be given antibiotics (as this may increase antibiotic resistance). Other people argue that this could cause **unnecessary suffering** to the animals.

## Practice Questions

Q1 Briefly describe a method for testing the antibiotic resistance of bacteria.

Q2 Briefly describe one situation where scientific knowledge has affected the decision-making surrounding antibiotic resistance.

**Exam Questions**

Q1 Put forward arguments for and against giving antibiotics to someone suffering from dementia. [2 marks]

Q2 A mother takes her son, who is suffering from a mild chest infection, to the doctor to get some antibiotics.

a) Why might the doctor be reluctant to prescribe the child antibiotics? [2 marks]
b) Why might the mother disagree if the doctor refuses to prescribe antibiotics? [2 marks]

Q3 A study was carried out to determine if the increase in the national rate of bacterial resistance to antibiotic X is linked to people not finishing their course of the antibiotic. Part of the study involved sending out questionnaires to 300 patients from one GP surgery in East Anglia.

Evaluate the methodology of this study. [3 marks]

## R-E-S-I-S-T-A-N-T — find out what it means to me...

*You're probably a bit bored of me ramming it down your throat now but you need to be able to evaluate any data or study you're presented with. Remember to look at the methodology, evidence and conclusions, and look out for anywhere there may be problems. And if you're ever asked to consider any ethical issues, think of the arguments for and against.*

# Human Impacts on Diversity

*Human activity has an impact on species diversity — and I'm not just talking about stepping on bugs. There are plenty of examples, but you need to know how agriculture and deforestation affect diversity...*

## Species Diversity is the Number of Species Present in a Community

1) Species diversity is the number of **different** species and the **abundance** of each species within a **community**.
2) The **higher** the species diversity of **plants and trees** in an area, the **higher** the species diversity of **insects, animals** and **birds**. This is because there are **more habitats** (places to live) and a larger and more varied **food source**.
3) Diversity can be **measured** to help us monitor ecosystems and identify areas where it has been **dramatically reduced**.

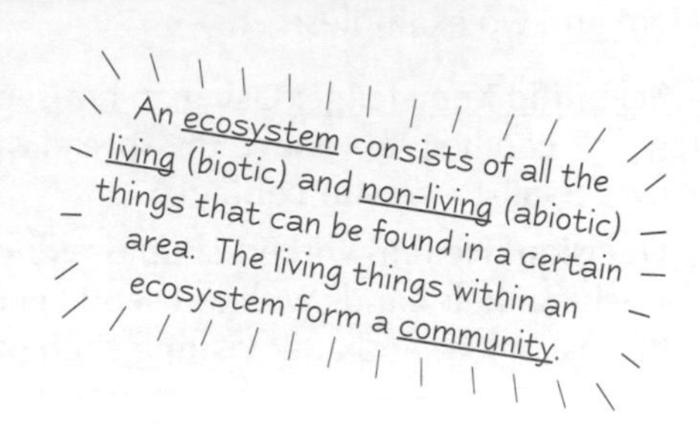


## Species Diversity is Measured using the Index of Diversity

1) The simplest way to measure diversity is just to **count up** the number of **different species**.
2) But that **doesn't** take into account the **population size** of each species.
3) Species that are in a community in very **small** numbers shouldn't be treated the same as those with **bigger** populations.
4) The **index of diversity** is calculated using an equation that takes different population sizes into account. You calculate the index of diversity (**d**) of a community using this formula:

$$d = \frac{N(N-1)}{\sum n(n-1)}$$

Where...
N = **Total number** of organisms of **all** species
n = **Total number** of **one** species
$\Sigma$ = '**Sum of**' (i.e. added together)

The **higher** the number the **more diverse** the area is. If all the individuals are of the same species (i.e. no diversity) the diversity index is 1. Here's an example:

There are 3 different species of flower in this field — a red species, a white and a blue.
There are 11 organisms altogether, so N = 11.
There are 3 of the red species, 5 of the white and 3 of the blue.
So the species diversity index of this field is:

$$d = \frac{11\,(11-1)}{3\,(3-1) + 5\,(5-1) + 3\,(3-1)} = \frac{110}{6 + 20 + 6} = 3.44$$

When calculating the bottom half of the equation you need to work out the n(n–1) bit for each different species then add them all together.

## Deforestation Decreases Species Diversity...

We cut down forests to get **wood** and **create land** for **farming** and **settlements**.
Here are some reasons why this affects diversity:

1) Deforestation **directly** reduces the **number** of **trees** and sometimes the **number** of **different tree species**.
2) Deforestation also **destroys habitats**, so some species could lose their **shelter** and **food source**. This means that these species will **die** or be forced to **migrate** to another suitable area, further **reducing** diversity.
3) The migration of organisms into increasingly smaller areas of remaining forest may **temporarily increase species diversity** in those areas.

# Human Impacts on Diversity

## ...as Does Agriculture

Farmers try to **maximise** the **amount of food** that they can produce from a given area of land. Many of the methods they use reduce diversity.

1) **Woodland clearance** — this is done to **increase** the **area** of farmland. This **reduces** species diversity for the same reasons as **deforestation** (see previous page).
2) **Hedgerow removal** — this is also done to **increase** the **area** of farmland by turning **lots of small fields** into **fewer large fields**. This **reduces** species diversity for the same reasons as **woodland clearance** and **deforestation**.
3) **Monoculture** — this is when farmers grow fields containing only **one type of plant**. A **single type** of plant will support **fewer species**, so diversity is **reduced**.
4) **Pesticides** — these are chemicals that **kill** organisms (**pests**) that feed on **crops**. This **reduces** diversity by **directly killing** the pests. Also, any species that feed on the pests will **lose** a food source, so their numbers could **decrease** too.
5) **Herbicides** — these are chemicals that kill **unwanted plants** (**weeds**). This **reduces** plant diversity and could **reduce** the number of organisms that feed on the weeds.

Pete wasn't sure that the company's new increased diversity policy would be good for productivity.

## Practice Questions

Q1 Why does greater plant and tree diversity increase insect, animal and bird diversity?

Q2 How is species diversity calculated?

Q3 Why does deforestation lead to reduced species diversity?

Q4 What is monoculture and how does it reduce species diversity?

Q5 Why do pesticides and herbicides reduce species diversity?

**Exam Question**

Q1 A study was conducted to investigate the impact of introducing enhanced field margins on the diversity of bumblebees. Enhanced field margins are thick bands of land around the edges of fields that are not farmed, but instead are planted with plants that are good for wildlife. Scientists studied two wheat fields, one where the farmer sowed crops right to the edge of the field and another where the farmer created enhanced field margins.

a) What is species diversity a measure of? [2 marks]

b) Use the data below to calculate the index of diversity for each site.

| Site 1 — No Field Margins | | Site 2 — Enhanced Field Margins | |
|---|---|---|---|
| *Bombus lucorum* | 15 | *Bombus lucorum* | 35 |
| *Bombus lapidarius* | 12 | *Bombus lapidarius* | 25 |
| *Bombus pascuorum* | 24 | *Bombus pascuorum* | 34 |
| | | *Bombus ruderatus* | 12 |
| | | *Bombus terrestris* | 26 |

[4 marks]

c) What conclusions can be drawn from the findings of this study? [2 marks]

## TIM-BER! — deforestation brings species diversity crashing down...

*There's nothing special about species diversity — just remember it's the number of species in a community and the abundance of each. Simple. Now for the maths bit... practise the index of diversity equation until you can't imagine life without it, then when it comes up in the exam it'll be a breeze — a nice warm one, not a spine-chilling one.*

# Interpreting Diversity Data

*I know it seems like there are lots of risks associated with human activity, but there are plenty of benefits too — we don't just cut trees down for the fun of it you know.*

## *Human Activity in an Area Has Benefits and Risks*

**Deforestation**

**Benefits**

1) **Wood** and **land** for homes to be built.
2) Local areas become more **developed** by attracting businesses.

**Risks**

1) **Diversity** is **reduced** — species could become **extinct**.
2) Less **carbon dioxide** is stored because there are fewer plants and trees, which contributes to **climate change**.
3) Many **medicines** come from organisms found in rainforests — possible future discoveries are **lost**.
4) **Natural beauty** is **lost**.

**Agriculture**

**Benefits**

1) **More food** can be produced.
2) Food is **cheaper** to produce, so food **prices** are **lower**.
3) Local areas become more **developed** by attracting businesses.

**Risks**

1) **Diversity** is **reduced** — because of **monoculture**, **woodland and hedgerow clearance, herbicide** and **pesticide use** (see previous page).
2) **Natural beauty** is **lost**.

## *You Might Have to Interpret Data About How Human Activity Affects Diversity*

You might have to interpret some data in the exam. Here are three examples of the kind of thing you might get:

**Example 1 — Herbicides**

Herbicides kill **unwanted plants** (weeds) whilst leaving **crops unharmed**.
The crops can then **grow better** because they're not **competing for resources** with weeds.
The graph below shows plant diversity in an **untreated** field and a field **treated annually** with **herbicide**.

**Describe the data:**

- Plant diversity in the **untreated field** showed a **slight increase** in the seven years. Plant diversity **decreased** a lot in the **treated field** when the **herbicide** was **first applied**. The diversity then **recovered** throughout each year, but was **reduced again** by **each** annual application of herbicide.

**Explain the data:**

- When the herbicide was applied, the weeds were **killed, reducing species diversity**. **In between** applications, diversity **increased** as **new weeds grew**. These were then killed again by each annual application of herbicide.

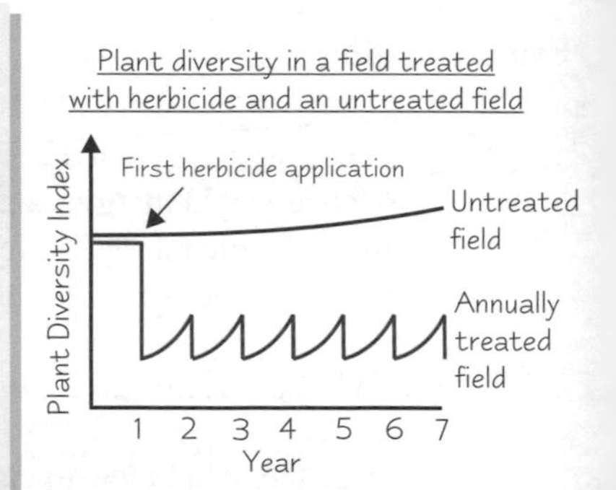


**Example 2 — Skylark Population**

Since the 1970s, farmers have been planting many crops in the **winter** instead of the **spring** to **maximise production**. This means there are **fewer** fields left as **stubble** over the winter. This is a problem because **skylarks** like to **nest** in fields with **stubble**. The graph below shows how the skylark population has **changed** in the UK since 1970.

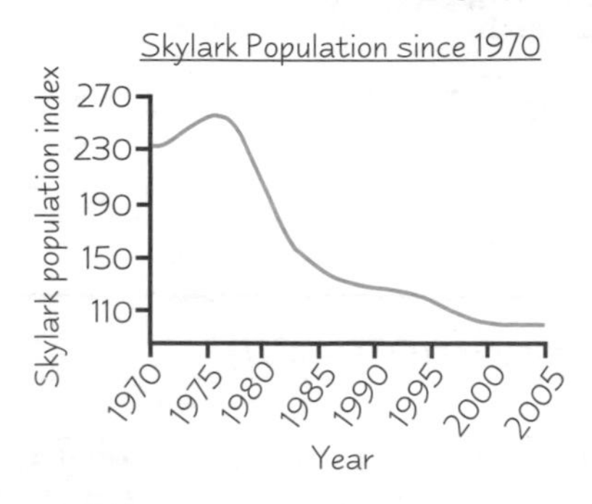


**Describe the data:**

- Skylark diversity showed a **small increase** from **1970** to the **late 1970s**. Since the late 1970s skylark diversity has **decreased** a lot. The skylark diversity has remained roughly **constant** from **2000** to **2005**.

**Explain the data:**

- With **fewer nesting sites** available, **fewer offspring** could be successfully raised, leading to a reduction in the number of skylarks.

# Interpreting Diversity Data

**Example 3 — Loss of Rainforest**

The graph below shows the results of a study that **compared** the **diversity** in a **rainforest** with the diversity in a **deforested area** that had been **cleared** for agricultural use.

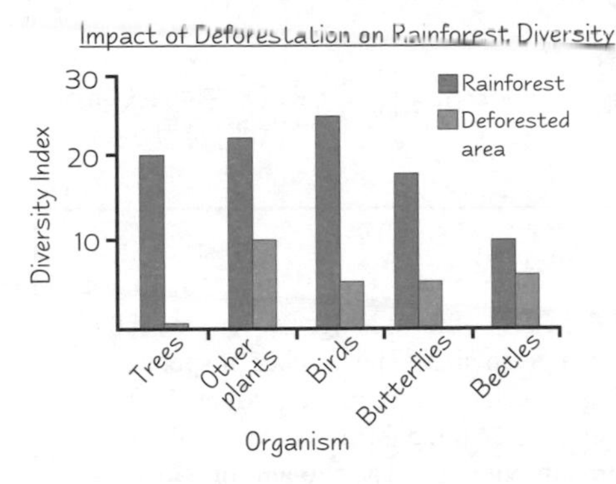


**Describe the data:**

- For **all types** of organism studied, **species diversity** is **higher** in the **rainforest** than in the **deforested area**.
- **Deforestation** has **reduced** the species diversity of **trees** the most.

**Explain the data:**

- Many organisms **can't adapt** to the **change in habitat** and must **migrate** or **die** — **reducing** diversity in the area.
- Reduced **tree diversity** leads to a reduction in the diversity of **all other organisms** (see p. 86).

## *Society Uses Diversity Data to Make Decisions*

**Diversity data** can be used to see which **species or areas** are being **affected** by **human activity**.
This information can then be used by **society** to **make decisions** about human activities. For example:

| Scientific Finding | Decision Made |
|---|---|
| Fewer hedgerows reduces diversity. | The UK government offers farmers money to encourage them to plant hedgerows, and to cover the cost of not growing crops on these areas. |
| Deforestation reduces diversity. | Some governments encourage sustainable logging (a few trees are taken from lots of different areas and young trees are planted to replace them). |
| Human development reduces diversity. | Many governments are setting up protected areas (e.g. national parks) where human development is restricted to help conserve diversity. |
| Some species are facing extinction. | Breeding programmes in zoos help to increase the numbers of endangered species in a safe environment before reintroducing them to the wild. |

## *Practice Questions*

Q1 Describe one risk associated with deforestation.

Q2 Describe one benefit of agricultural activities.

**Exam Question**

Q1 The graph on the right shows the results from a study conducted into wild bird populations in the UK between 1970 and 2006. It shows the pattern of change for woodland and farmland species.

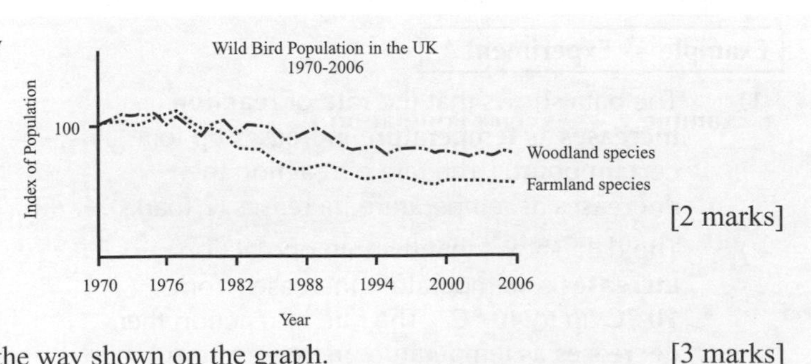


a) Describe the data. [2 marks]

b) Human activity has significantly affected wild bird populations. Suggest reasons why the woodland and farmland species have changed in the way shown on the graph. [3 marks]

c) Discuss the potential benefits of agriculture and deforestation and the associated risks to diversity. [8 marks]

## *Diver-city — it's a wonderful place where you get to jump off stuff...*

*Interpreting data is just about understanding what the graphs or tables are telling us. Describing it is simple — just say what you see. Then you've usually got to explain it — say what the reasons might be for what you've described. After that they might ask you to do a little song, or a little dance, or maybe jump through a couple of hoops...*

# How to Interpret Experiment and Study Data

*Science is all about getting good evidence to test your theories... so scientists need to be able to spot a badly designed experiment or study a mile off, and be able to interpret the results of an experiment or study properly. Being the cheeky little monkeys they are, your exam board will want to make sure you can do it too. Here's a quick reference section to show you how to go about interpreting data-style questions.*

## Here Are Some **Things** You Might be **Asked** to do...

For other examples check the interpreting data pages in the sections.

Here are two examples of the kind of data you could expect to get:

**Experiment A**

Experiment A examined the effect of temperature on the rate of an enzyme-controlled reaction. The rate of reaction for enzyme X was measured at six different temperatures (from 10 to 60 °C). All other variables were kept constant. A negative control containing all solutions except the enzyme was included. The rate of reaction for the negative control was zero at each temperature used. The results are shown in the graph below.

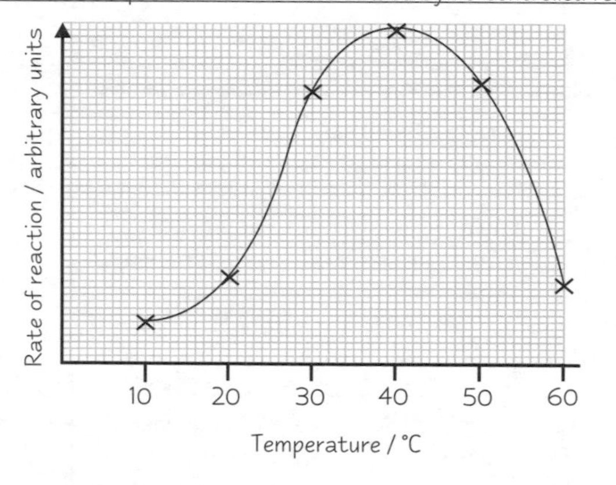


**Study B**

Study B examined the effect of farm hedgerow length on the number of species in a given area. The number of species present during a single week on 12 farms was counted by placing ground-level traps. All the farms were a similar area. The traps were left out every day, at 6 am for two hours and once again at 6 pm for two hours. The data was plotted against hedgerow length. The results are shown in the scattergram below.

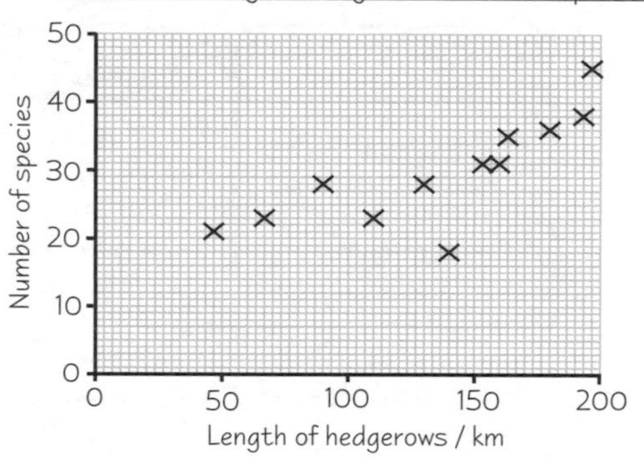


## 1) **Describe** the **Data**

You need to be able to **describe** any data you're given. The level of **detail** in your answer should be appropriate for the **number of marks** given. Loads of marks = more detail, few marks = less detail.
For the two examples above:

**Example — Experiment A**

1) The data shows that the **rate of reaction increases** as **temperature increases** up to a **certain point**. The rate of reaction then **decreases** as temperature increases (2 marks).
2) The data shows that the rate of reaction **increases** as temperature increases from **10 °C up to 40 °C**. The rate of reaction then **decreases** as temperature increases from **40 °C to 60 °C** (4 marks).

**Example — Study B**

The data shows a **positive correlation** between the length of hedgerows and the number of species in the area (1 mark).

Correlation describes the **relationship** between two variables — the one that's been changed and the one that's been measured. Data can show **three** types of correlation:

1) **Positive** — as one variable **increases** the other **increases**.
2) **Negative** — as one variable **increases** the other **decreases**.
3) **None** — there is **no relationship** between the two variables.

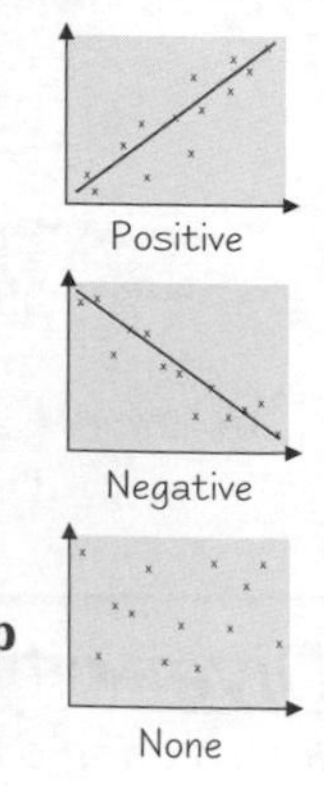

# How to Interpret Experiment and Study Data

## 2) *Draw* or *Check* the *Conclusions*

1) Ideally, only **two** quantities would ever change in any experiment or study — everything else would be **constant**.
2) If you can keep everything else constant and the results show a correlation then you **can** conclude that the change in one variable **does cause** the change in the other. →

**Example — Experiment A**

All other variables were **kept constant**. E.g. pH, enzyme concentration and substrate concentration **stayed the same** each time, so these **couldn't** have influenced the change in the rate of reaction. So you **can say** that an increase in temperature **causes** an increase in the rate of reaction up to a certain point.

3) But usually all the variables **can't** be controlled, so other **factors** (that you **couldn't** keep constant) could be having an **effect**.
4) Because of this, scientists have to be very careful when **drawing conclusions**. Most results show a **link** (correlation) between the variables, but that **doesn't prove that a change in one causes the change in the other**. →

**Example — Study B**

The length of hedgerows shows a **positive correlation** with the number of species in that area. But you **can't** conclude that fewer hedgerows **causes** fewer species. **Other factors** may have been involved, e.g. the number of **predators** of the species studied may have increased in some areas, the farmers may have used **more pesticide** in one area, or something else you hadn't thought of could have caused the pattern...

5) The **data** should always **support** the conclusion. This may sound obvious but it's easy to **jump** to conclusions. Conclusions have to be **precise** — not make sweeping generalisations. →

**Example — Experiment A**

A science magazine **concluded** from this data that enzyme X works best at **40 °C**. The data **doesn't** support this. The enzyme **could** work best at 42 °C, or 47 °C but you can't tell from the data because **increases** of **10 °C** at a time were used. The rates of reaction at in-between temperatures **weren't** measured.

## 3) *Comment on the* Reliability *of the* Results

**Reliable** means the results can be **consistently reproduced** in independent experiments. And if the results are reproducible they're more likely to be **true**. If the data isn't reliable for whatever reason you **can't draw** a valid **conclusion**. Here are some of the things that affect the reliability of data:

Davina wasn't sure she'd got a large enough sample size.

1) __Size of the data set__ — For experiments, the **more repeats** you do, the **more reliable** the data. If you get the **same result** twice, it could be the correct answer. But if you get the same result **20 times**, it's much more reliable. The general rule for **studies** is the **larger** the sample size, the more **reliable** the **data** is.

   E.g. Study B is quite **small** — they only used 12 farms. The **trend** shown by the data may not appear if you studied **50 or 100 farms**, or studied them for a longer period of time.

2) __Variables__ — The **more variables** you **control**, the **more reliable** your data is. In an experiment you would control all the variables, but when doing a study this isn't always possible. You try to control **as many as possible** or use **matched groups** (see page 3).

   E.g. ideally, all the farms in Study B would have a similar **type** of land, similar **weather**, have the same **crops** growing, etc. Then you could be more sure that the one factor being **investigated** (hedgerows) is having an **effect** on the thing being **measured** (number of species). In Experiment A, **all** other variables were controlled, e.g. pH, concentrations, volumes, so you can be sure the temperature is causing the **change** in the **reaction rate**.

Jane rarely ate chocolate, honestly.

3) __Data collection__ — think about all the **problems** with the **method** and see if **bias** has slipped in. For example, members of the public sometimes tell **little porkies**, so it's easy for studies involving **questionnaires** to be **biased**. E.g. people often underestimate how much alcohol they drink or how many cigarettes they smoke.

   E.g. in Study B, the traps were placed on the **ground**, so species like birds weren't included. The traps weren't left overnight, so **nocturnal** animals wouldn't get counted, etc. This could have affected the results.

# How to Interpret Experiment and Study Data

4) **Controls** — without controls, it's very difficult to **draw valid conclusions**. **Negative controls** are used to make sure that nothing you're doing in the experiment has an effect, **other than** what you're testing. But it's not always possible to have controls in studies (study controls usually involve a group where **nothing changes**, e.g. a group of patients aren't given a new long-term treatment to make sure any effects detected in the patients having the treatment aren't due to the fact that they've had two months to recover).

E.g. in Experiment A, the **negative control** contained everything from the experiment **except** the enzyme. This was used to show that the change in reaction rate was caused by the effect of **temperature** on the **enzyme**, and nothing else. If something else in the experiment (e.g. the water, or something in the test tube) was causing the change, you would get the **same results** in the negative control (and you'd know something was up).

5) **Repetition by other scientists** — for theories to become accepted as 'fact' other scientists need to **repeat** the work (see page 2). If **multiple studies** or **experiments** come to the same conclusion, then that conclusion is **more reliable**.

E.g. if a second group of scientists carried out the same experiment for enzyme X and got the same results, the results would be **more reliable**.

## 4) Analyse the Data

Sometimes it's easier to **compare data** by making a few calculations first, e.g. converting raw data into **ratios** or **percentages**.

Example Three UK hospitals have been trying out three **different methods** to **control the spread** of chest infections. A study investigated the number of people suffering from chest infections in those hospitals over a **three month period**. The table opposite shows the results. If you just look at the **number of cases** in the **last month** (March) then the method of hospital 3 appears to have worked **least well**, as they have the **highest number** of infections. But if you look at the **percentage increase** in infections you get a different picture: hospital 1 = 30%, hospital 2 = 293%, and hospital 3 = 18%. So hospital 3 has the lowest percentage increase, suggesting their method of control is **working the best**.

| | Number of cases per 6000 patients | | |
|---|---|---|---|
| Hospital | Jan | Feb | March |
| 1 | 60 | 65 | 78 |
| 2 | 14 | 24 | 55 |
| 3 | 93 | 96 | 110 |

Calculating percentage increase, hospital 1:

$$\frac{(78 - 60)}{60} \times 100 = \frac{18}{60} \times 100 = 30\%$$

## There Are a Few Technical Terms You Need to Understand

I'm sure you probably know these all off by heart, but it's easy to get mixed up sometimes. So here's a quick recap of some words **commonly used** when assessing and analysing experiments and studies:

1) **Variable** — A variable is a **quantity** that has the **potential to change**, e.g. weight. There are two types of variable commonly referred to in experiments:
   - **Independent variable** — the thing that's **changed** in an experiment.
   - **Dependent variable** — the thing that you **measure** in an experiment.

When drawing graphs, the dependent variable should go on the **y-axis** (the vertical axis) and the independent on the **x-axis** (the horizontal axis).

2) **Accurate** — Accurate results are those that are **really close** to the **true** answer.

3) **Precise results** — These are results taken using **sensitive instruments** that measure in **small increments**, e.g. pH measured with a meter (pH 7.692) will be **more precise** than pH measured with paper (pH 8).

It's possible for results to be precise **but not** accurate, e.g. a balance that weighs to 1/1000 th of a gram will give precise results, but if it's not **calibrated** properly the results won't be accurate.

4) **Qualitative** — A **qualitative** test tells you **what's** present, e.g. an acid or an alkali.

5) **Quantitative** — A **quantitative** test tells you **how much** is present, e.g. an acid that's pH 2.46.

## Controls — I think I prefer the remote kind...

*These pages should give you a fair idea of the points to think about when interpreting data. Just use your head and remember the three main points in the checklist — **d**escribe the **d**ata, **c**heck the **c**onclusions and make sure the **r**esults are **r**eliable.*

# Answers

## Unit 1: Section 1 — Disease and Immunity

### Page 5 — Disease

1 *Maximum of 3 marks available.*
*A pathogen may rupture the host cells* ***[1 mark]****. It may break down and use nutrients in the host cells, so that the host cells starve* ***[1 mark]****. Or a pathogen may replicate inside host cells and burst them as it leaves* ***[1 mark]****.*

2 *Maximum of 2 marks available.*
*Axes correct way round with time spent sunbathing on x-axis and incidence of skin cancer on y-axis* ***[1 mark]****. Positive correlation between two variables* ***[1 mark]****.*
*E.g.*

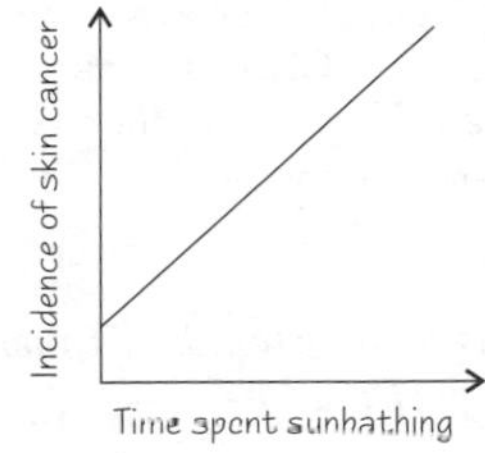


### Page 7 — The Immune System

1 *Maximum of 3 marks available.*
*Antibodies coat pathogens, making it easier for phagocytes to engulf them* ***[1 mark]*** *and preventing them from entering host cells* ***[1 mark]****. They also bind to toxins to neutralise them* ***[1 mark]****.*
There are three marks available for this question, so you need to think of three different functions.

2 *Maximum of 6 marks available.*
*A secondary immune response is faster* ***[1 mark]*** *and produces a quicker, stronger response* ***[1 mark]*** *than the primary response. This is because memory cells are produced during the primary response* ***[1 mark]*** *which remember the foreign antigen* ***[1 mark]****. During the second infection, memory cell B-cells can quickly divide to form plasma cells, which secrete the correct antibody to the antigen* ***[1 mark]****. Memory T-cells quickly divide into the right type of T-cells to kill the cell carrying the antigen* ***[1 mark]****.*
You'll only get the full marks for this question if you explain (as well as describe) why the secondary response differs.

### Page 9 — Vaccines and Antibodies in Medicine

1 *Maximum of 4 marks available.*
*The flu virus is able to change its surface antigens/shows antigenic variation* ***[1 mark]****. This means that when you're infected for a second time with a different strain, the memory cells produced from the first infection will not recognise the new/different antigens* ***[1 mark]****. The immune system has to carry out a primary response against these new antigens* ***[1 mark]****. This takes time and means you become ill* ***[1 mark]****.*

2 *Maximum of 4 marks available.*
*Monoclonal antibodies are made against antigens specific to cancer cells* ***[1 mark]****. An anti-cancer drug is attached to the antibodies* ***[1 mark]****. The antibodies bind to tumour markers on cancer cells because their binding sites have a complementary shape* ***[1 mark]****. This delivers the anti-cancer drug to the cells* ***[1 mark]****.*

### Page 11 — Interpreting Vaccine and Antibody Data

1 a) *Maximum of 2 marks available.*
*Because people were immunised against Hib* ***[1 mark]*** *and also had the protection of herd immunity* ***[1 mark]****.*
b) *Maximum of 1 mark available.*
*The number of cases of Hib increased* ***[1 mark]****.*

## Unit 1: Section 2 — The Digestive System

### Page 13 — The Digestive System

1 a) *Maximum of 6 marks available.*
*The pancreas releases pancreatic juice into the small intestine/duodenum* ***[1 mark]****. Pancreatic juice contains amylase, trypsin, chymotrypsin and lipase* ***[1 mark]****. Amylase breaks down starch into maltose* ***[1 mark]****. Chymotrypsin and trypsin break down proteins into peptides* ***[1 mark]****. Lipase breaks down lipids into fatty acids and glycerol* ***[1 mark]****. Pancreatic juice also neutralises acid from the stomach* ***[1 mark]****.*
b) *Maximum of 1 mark available.*
*E.g. salivary glands* ***[1 mark]***

### Page 15 — Proteins

1 *Maximum of 5 marks available*
*Two amino acids join together in a condensation reaction* ***[1 mark]****. A peptide bond* ***[1 mark]*** *forms between the carboxyl group* ***[1 mark]*** *of one amino acid and the amino group* ***[1 mark]*** *of the other amino acid. A molecule of water is released* ***[1 mark]****.*
If you find it difficult to explain a process, such as a dipeptide forming, learn the diagrams too because they may help you to explain the process.

2 *Maximum of 9 marks available.*
*Proteins are made from amino acids* ***[1 mark]****. The amino acids are joined together in a long (polypeptide) chain* ***[1 mark]****. The sequence of amino acids is the protein's primary structure* ***[1 mark]****. The amino acid chain/polypeptide coils or folds in a certain way* ***[1 mark]****. The way it's coiled or folded is the protein's secondary structure* ***[1 mark]****. The coiled or folded chain is itself coiled or folded into a specific shape* ***[1 mark]****. This is the protein's tertiary structure* ***[1 mark]****. Different polypeptide chains can be joined together in the protein molecule* ***[1 mark]****. The way these chains are joined together is the quaternary structure of the protein* ***[1 mark]****.*
The question specifically states that you don't need to describe the chemical nature of the bonds in a protein. So, even if you name them, don't go into chemical details of how they're formed — no credit will be given.

### Page 17 — Carbohydrates

1 *Maximum of 4 marks available*
*Lactose intolerance is caused by a lack of the enzyme lactase* ***[1 mark]****. Sufferers don't have enough lactase to break down lactose, a sugar found in milk/milk products* ***[1 mark]****. Undigested lactose is fermented by bacteria* ***[1 mark]****. This can lead to intestinal complaints such as stomach cramps, flatulence and diarrhoea* ***[1 mark]****.*

# Answers

2 *Maximum of 6 marks available*
*Add dilute hydrochloric acid to the solution and boil* ***[1 mark]****.*
*Neutralise with sodium hydrogencarbonate* ***[1 mark]****.*
*Add blue Benedict's reagent to the solution and heat* ***[1 mark]****.*
*If a brick red precipitate forms this indicates that either non-reducing, or reducing sugars are present* ***[1 mark]****.*
*Carry out the reducing sugar test* ***[1 mark]*** *(heating with Benedict's reagent* ***[1 mark]****) to determine which of the two sugars are present.*
The question asks to describe a test for a non-reducing sugar. Remember, there are a couple more steps involved in the test for a non-reducing sugar than in the test for a reducing sugar.

## Page 19 — Enzyme Action

1 *Maximum of 7 marks available.*
*In the 'lock and key' model the enzyme and the substrate have to fit together at the active site of the enzyme* ***[1 mark]****.*
*This creates an enzyme-substrate complex* ***[1 mark]****.*
*The active site then causes changes in the substrate* ***[1 mark]****.*
This mark could also be gained by explaining the change (e.g. bringing molecules closer together, or putting a strain on bonds).
*The change results in the substrate being broken down/joined together* ***[1 mark]****.*
*The 'induced fit' model has the same basic mechanism as the 'lock and key' model* ***[1 mark]****.*
*The difference is that the substrate is thought to cause a change in the enzyme's active site shape* ***[1 mark]****, which enables a better fit* ***[1 mark]****.*

2 *Maximum of 2 marks available.*
*A change in the amino acid sequence of an enzyme may alter its tertiary structure* ***[1 mark]****. This changes the shape of the active site so that the substrate can't bind to it* ***[1 mark]****.*

## Page 21 — Factors Affecting Enzyme Activity

1 *Maximum of 8 marks available, from any of the 10 points below.*
*If the solution is too cold, the enzyme will work very slowly* ***[1 mark]****. This is because, at low temperatures, the molecules have little kinetic energy, so move slowly, making collisions between enzyme and substrate molecules less likely* ***[1 mark]****.*
*Also, fewer of the collisions will have enough energy to result in a reaction* ***[1 mark]****.*
The marks above could also be obtained by giving the reverse argument — a higher temperature is best to use because the molecules will move fast enough to give a reasonable chance of collisions and those collisions will have more energy, so more will result in a reaction.
*If the temperature gets too high, the reaction will stop* ***[1 mark]****.*
*This is because the enzyme is denatured* ***[1 mark]*** *— the active site changes shape and will no longer fit the substrate* ***[1 mark]****.*
*Denaturation is caused by increased vibration breaking bonds in the enzyme* ***[1 mark]****. Enzymes have an optimum pH* ***[1 mark]****.*
*pH values too far from the optimum cause denaturation* ***[1 mark]****.*
Explanation of denaturation here will get a mark only if it hasn't been explained earlier.
*Denaturation by pH is caused by disruption of ionic and hydrogen bonds, which alters the enzyme's tertiary structure* ***[1 mark]****.*

2 a) *Maximum of 3 marks available.*
*Competitive inhibitor molecules have a similar shape to the substrate molecules* ***[1 mark]****. They compete with the substrate molecules to bind to the active site of an enzyme* ***[1 mark]****. When an inhibitor molecule is bound to the active site it stops the substrate molecule from binding* ***[1 mark]****.*

b) *Maximum of 2 marks available.*
*Non-competitive inhibitor molecules bind to enzymes away from their active site* ***[1 mark]****. This causes the active site to change shape so the substrate molecule can no longer fit* ***[1 mark]****.*

# Unit 1: Section 3 — Cell Structure and Membranes

## Page 23 — Animal Cell Structure

1 *Maximum of 4 marks available.*
*ribosomes* ***[1 mark]****, rough endoplasmic reticulum* ***[1 mark]****, Golgi apparatus* ***[1 mark]****, plasma membrane* ***[1 mark]***
This question really tests how well you know what each organelle does. The rough endoplasmic reticulum transports proteins that have been made in the ribosomes to the Golgi apparatus. At the Golgi apparatus the proteins are packaged and sent to the plasma membrane to be secreted or inserted in the membrane itself.

2 *Maximum of 2 marks available.*
*Ciliated epithelial cells have lots of mitochondria* ***[1 mark]*** *because they need lots of energy* ***[1 mark]****.*

## Page 25 — Analysis of Cell Components

1 *Maximum of 6 marks available.*
*TEMs use electromagnets to focus a beam of electrons, which is transmitted through the specimen* ***[1 mark]****. Denser parts of the specimen absorb more electrons and appear darker* ***[1 mark]****. SEMs scan a beam of electrons across the specimen* ***[1 mark]****. This knocks off electrons from the specimen, which are gathered in a cathode ray tube, to form an image* ***[1 mark]****. TEMs can only be used on thin specimens* ***[1 mark]****. SEMs produce lower resolution images than TEMs* ***[1 mark]****.*

2 *Maximum of 8 marks available.*
*First, the cell sample is homogenised* ***[1 mark]*** *to break up the plasma membranes and release the organelles into solution* ***[1 mark]****. The cell solution is then filtered* ***[1 mark]*** *to remove any large cell debris or tissue debris* ***[1 mark]****. Next the solution is ultracentrifuged* ***[1 mark]*** *to separate out the different types of organelles* ***[1 mark]****. The organelles are separated according to mass, with the heaviest being separated first* ***[1 mark]****. Centrifugation is repeated at higher and higher speeds to separate out the lighter and lighter organelles* ***[1 mark]****.*
Make sure you remember to explain each step otherwise you won't be able to get full marks.

## Page 27 — Plasma Membranes

1 *Maximum of 2 marks available.*
*The membrane is described as fluid because the phospholipids are constantly moving* ***[1 mark]****. It is described as a mosaic because the proteins are scattered throughout the membrane like tiles in a mosaic* ***[1 mark]****.*

2 a) *Maximum of 3 marks available.*
*Two fatty acid molecules* ***[1 mark]*** *and a phosphate group* ***[1 mark]*** *attached to one glycerol molecule* ***[1 mark]****.*
Don't get phospholipids mixed up with triglycerides — a triglyceride has three fatty acids attached to one glycerol molecule.

b) *Maximum of 2 marks available.*
*Saturated fatty acids don't have any double bonds between their carbon atoms* ***[1 mark]****. Unsaturated fatty acids have one or more double bonds between their carbon atoms* ***[1 mark]****.*

# Answers

## Page 29 — Exchange Across Plasma Membranes

1 a) *Maximum of 3 marks available.*
*The water potential of the sucrose solution was higher than the water potential of the potato **[1 mark]**. Water moves by osmosis from a solution of higher water potential to a solution of lower water potential **[1 mark]**. So water moved into the potato, increasing its mass **[1 mark]**.*

b) *Maximum of 1 mark available.*
*The water potential of the potato and the water potential of the solution was the same **[1 mark]**.*

c) *Maximum of 4 marks available.*
*– 0.4 g **[1 mark]**. The potato has a higher water potential than the solution **[1 mark]** so net movement of water is out of the potato **[1 mark]**. The difference in water potential between the solution and the potato is the same as with the 1% solution, so the mass difference should be about the same **[1 mark]**.*

## Page 31 — Exchange Across Plasma Membranes

1 *Maximum of 10 marks available.*
*Some glucose moves across the intestinal epithelial cells and into the blood by diffusion **[1 mark]**. This is because initially there is a higher concentration of glucose in the lumen of the small intestine than in the blood **[1 mark]**. The rest of the glucose moves into the blood by co-transport with sodium ions **[1 mark]**. There is a higher concentration of sodium ions in the lumen of the small intestine than inside the intestinal epithelial cell **[1 mark]**. This is because sodium ions are actively transported out of the cell into the blood by a sodium-potassium pump **[1 mark]**. So sodium ions diffuse from the small intestine lumen into the cell **[1 mark]** through sodium-glucose co-transporter proteins **[1 mark]**. These co-transporter proteins carry glucose into the cell along with the sodium **[1 mark]**. The concentration of glucose inside the cell increases **[1 mark]** and glucose diffuses out of the cell, into the blood **[1 mark]**.*
To answer this question it may help if you remember the diagram and then work through it step by step in your head, writing down each point. Also you need to remember the two methods of glucose transport (diffusion and co-transport) to get full marks.

2 *Maximum of 5 marks available.*
*The hydrophobic tails **[1 mark]** of the phospholipid bilayer prevent water-soluble molecules from diffusing through the membrane **[1 mark]**. Protein channels **[1 mark]** and carrier proteins **[1 mark]** control which of these water-soluble substances can enter and leave the cell **[1 mark]**.*

## Page 33 — Cholera

1 *Maximum of 5 marks available.*
*The bacterium produces a toxin **[1 mark]** that causes chloride channels in the lining of the small intestine to open **[1 mark]**. Chloride ions diffuse into the small intestine **[1 mark]**. The small intestine now has a lower water potential than the blood **[1 mark]**, so water moves from the blood into the small intestine, causing diarrhoea and dehydration **[1 mark]**.*

2 *Maximum of 2 marks available.*
*Against: e.g. children can't make their own decision to be part of the trial **[1 mark]**. For: e.g. scientists believe treatment for a disease that mainly affects children must be tested on children **[1 mark]**.*

# Unit 1: Section 4 — The Respiratory System

## Page 35 — Lung Function

1 *Maximum of 6 marks available.*
*There's a thin exchange surface **[1 mark]** as the alveolar epithelium is only one cell thick **[1 mark]**. This means there's a short diffusion pathway, which increases the rate of diffusion **[1 mark]**. The number of alveoli provide a large surface area for gas exchange, which also increases the rate of diffusion **[1 mark]**. There's a steep concentration gradient between the alveoli and the capillaries surrounding them, which increases the rate of diffusion **[1 mark]**. This is maintained by the flow of blood and ventilation **[1 mark]**.*

## Page 37 — How Lung Disease Affects Function

1 *Maximum of 7 marks available.*
*Fibrosis is the formation of scar tissue in the lungs **[1 mark]**. Scar tissue is less elastic than normal lung tissue, so lungs are less able to expand **[1 mark]** and tidal volume is reduced **[1 mark]**. It's also more difficult for gases to diffuse across the thicker scar tissue **[1 mark]** so less oxygen reaches the bloodstream **[1 mark]** and the rate of respiration in the cells is slower **[1 mark]**. Less energy is released, which results in the person feeling tired and weak **[1 mark]**.*

## Page 39 — Interpreting Lung Disease Data

1 a) *Maximum of 3 marks available.*
*The daily death rate increased rapidly after 4th December **[1 mark]** peaking around the 7th then decreasing afterwards **[1 mark]**. Both pollutants followed the same pattern **[1 mark]**.*
You could also get the marks by saying it the other way round — the pollutants rose and peaked around the 7th then decreased, with the death rates following the same pattern.

b) *Maximum of 1 mark available.*
*There is a link/correlation between the increase in sulfur dioxide and smoke concentration and the increase in death rate **[1 mark]**.*
Don't go saying that the increase in sulfur dioxide and smoke <u>caused</u> the increase in death rate — there could have been another reason for the trend, e.g. there could have been other pollutants responsible for the deaths.

# Unit 1: Section 5 — The Circulatory System

## Page 42 — The Heart

1 *Maximum of 6 marks available.*
*The valves only open one way **[1 mark]**. Whether they open or close depends on the relative pressure of the heart chambers **[1 mark]**. If the pressure is greater behind a valve (i.e. there's lots of blood in the chamber behind it) **[1 mark]**, it's forced open, to let the blood travel in the right direction **[1 mark]**. Once the blood has gone through the valve, the pressure is greater in front of the valve **[1 mark]**, which forces it shut, preventing blood from flowing back into the chamber **[1 mark]**.*
Here you need to explain how valves function in relation to blood flow, rather than just in relation to relative pressures.

2 a) *Maximum of 1 mark available.*
*0.2 - 0.4 seconds **[1 mark]**.*
The AV valves are shut when the pressure is higher in the ventricles than in the atria.

b) *Maximum of 1 mark available.*
*0.3 - 0.4 seconds **[1 mark]**.*
When the ventricles relax the volume of the chamber increases and the pressure falls. The pressure in the left ventricle was 16.5 kPa at 0.3 seconds and it decreased to 7.0 kPa at 0.4 seconds, so it must have started to relax somewhere between these two times.

# Answers

## Page 45 — Cardiovascular Disease

1 *Maximum of 4 marks available*
*Atheromas can lead to the formation of a blood clot/thrombosis **[1 mark]**. A blood clot could block blood flow to the heart muscle **[1 mark]**, causing a lack of oxygen, which damages the heart muscle **[1 mark]** and can lead to a heart attack **[1 mark]**.*
Be specific with the wording of your answers. Examiners won't award marks for unscientific phrases such as "putting strain on the heart".

2 *Maximum of 3 marks available.*
*Atheroma plaques damage and weaken arteries **[1 mark]** and can lead to increased blood pressure **[1 mark]**. When blood at high pressure travels through a weakened artery, the pressure can push the inner layers of the artery through the outer layer to form an aneurysm **[1 mark]**.*
This question is not asking about the consequences of an aneurysm, so no extra marks will be given if you write about it.

# Unit 2: Section 1 — Variation

## Page 47 — Causes of Variation

1 a) *Maximum of 3 marks available.*
*For species A, as the temperature increases the development time decreases **[1 mark]**. For species B the development time also decreases as the temperature increases **[1 mark]**. The development time of species B is less affected by temperature than species A **[1 mark]**.*

b) *Maximum of 4 marks available.*
*The variation between the species is mainly due to their different genes **[1 mark]**. Variation within a species is caused by both genetic and environmental factors **[1 mark]**. Individuals have different forms of the same genes (alleles), which causes genetic differences **[1 mark]**. Individuals may have the same genes, but environmental factors affect how they're expressed in their appearance (phenotype) **[1 mark]**.*

## Page 49 — Investigating Variation

1 a) *Maximum of 1 mark available.*
*To provide a control against which the women who smoked could be compared **[1 mark]**.*

b) *Maximum of 4 marks available.*
*Environmental factors (smoking) affect birth mass **[1 mark]**. Women who smoked showed a mean reduction in the birth mass of their babies of 377 g **[1 mark]**. Genetic factors also affect birth mass of babies born to women who smoke **[1 mark]**. The reduction in birth mass was as much as 1285 g among women who smoked and had certain genotypes **[1 mark]**.*

c) *Maximum of 2 marks available, from any of the 5 points below.*
*E.g. pre-pregnancy mass of the mothers **[1 mark]**, age **[1 mark]**, fitness levels **[1 mark]**, ethnic origin **[1 mark]**, if they'd had a previous pregnancy **[1 mark]**.*
Think of all the variables that need to be considered to isolate smoking as the only environmental factor that is influencing the variation.

# Unit 2: Section 2 — Genetics

## Page 51 — DNA

1 *Maximum of 4 marks available, from any of the 5 points below.*
*Nucleotides are joined between the phosphate group of one nucleotide and the sugar of the next **[1 mark]** forming the sugar-phosphate backbone **[1 mark]**. The two polynucleotide strands join through hydrogen bonds **[1 mark]** between the base pairs **[1 mark]**. The final mark is given for at least one accurate diagram showing at least one of the above points **[1 mark]**.*
As the question asks for a diagram make sure you do at least one, e.g.:

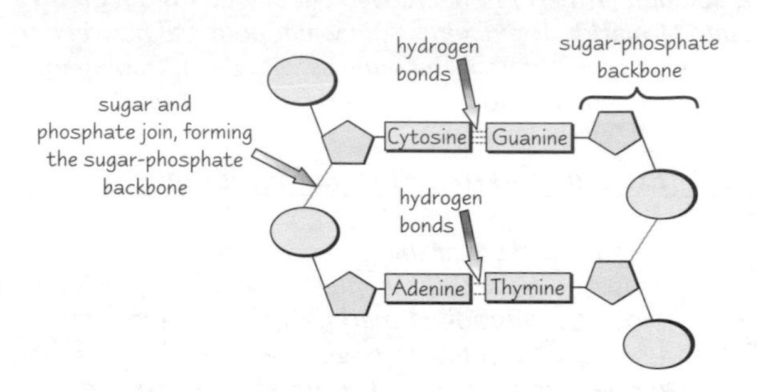


2 *Maximum of 5 marks available.*
*In eukaryotes, DNA is linear **[1 mark]** and wound around proteins (histones) **[1 mark]**.*
*In prokaryotes, DNA molecules are shorter **[1 mark]**, circular **[1 mark]** and not associated with proteins **[1 mark]**.*

## Page 53 — Genes

1 a) *Maximum of 2 marks available.*
*A gene is a section of DNA **[1 mark]** that codes for a protein (polypeptide) **[1 mark]**.*

b) *Maximum of 1 mark available.*
*Tryptophan, proline, proline, glutamic acid **[1 mark]**.*

2 *Maximum of 4 marks available.*
*The DNA sequence codes for the sequences of amino acids in proteins **[1 mark]**. Enzymes are proteins, so DNA codes for all enzymes **[1 mark]**. Enzymes control metabolic pathways **[1 mark]**. Metabolic pathways help to determine nature and development **[1 mark]**.*

## Page 55 — Meiosis and Genetic Variation

1 *Maximum of 2 marks available, from any of the 3 points below.*
*Normal body cells have two copies of each chromosome **[1 mark]**. Gametes have to have half the number of chromosomes so that when fertilisation takes place, the resulting embryo will have the correct diploid number **[1 mark]**. If the gametes had a diploid number, the resulting offspring would have twice the number of chromosomes that it should have **[1 mark]**.*

# Answers

2 Maximum of 6 marks available, from any of the 7 points below. The DNA unravels and replicates **[1 mark]**. The DNA condenses to form double-armed chromosomes **[1 mark]**. The chromosomes arrange themselves into homologous pairs **[1 mark]**. The pairs separate **[1 mark]**. The pairs of sister chromatids then separate **[1 mark]**. Four haploid, genetically different cells are produced **[1 mark]**. The final mark is given for at least one accurate diagram showing at least one of the above points **[1 mark]**.
As the question asks for a diagram make sure you do at least one, e.g.:

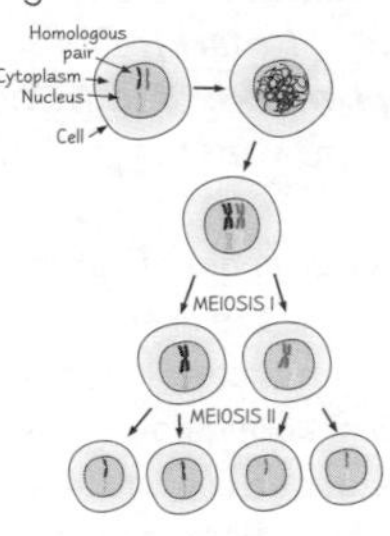


3 a) Maximum of 4 marks available.
During meiosis I, homologous pairs of chromosomes come together **[1 mark]**. The chromatids twist around each other and bits swap over **[1 mark]**. The chromatids now contain different combinations of alleles **[1 mark]**. This means each of the four daughter cells will contain chromatids with different combinations of alleles **[1 mark]**.

b) Maximum of 2 marks available.
Independent segregation means the chromosome pairs can split up in any way **[1 mark]**. So, the daughter cells produced can contain any combination of maternal and paternal chromosomes with different alleles **[1 mark]**.

## Page 57 — Genetic Diversity

1 Maximum of 3 marks available.
An event that causes a big reduction in a population, e.g. many members of a population die **[1 mark]**. A small number of members survive and reproduce **[1 mark]**. Because there are fewer members, there are fewer alleles in the new population, so the genetic diversity is reduced **[1 mark]**.

2 Maximum of 3 marks available.
Selective breeding involves humans selecting which organisms to breed until they produce one with the desired characteristics **[1 mark]**. Only organisms with similar traits and therefore similar alleles are bred together **[1 mark]**. So, the number of alleles in the population is reduced, resulting in reduced genetic diversity **[1 mark]**.

## Unit 2: Section 3 — Variation in Biochemistry and Cell Structure

### Page 59 — Variation in Haemoglobin

1 Maximum of 6 marks available.

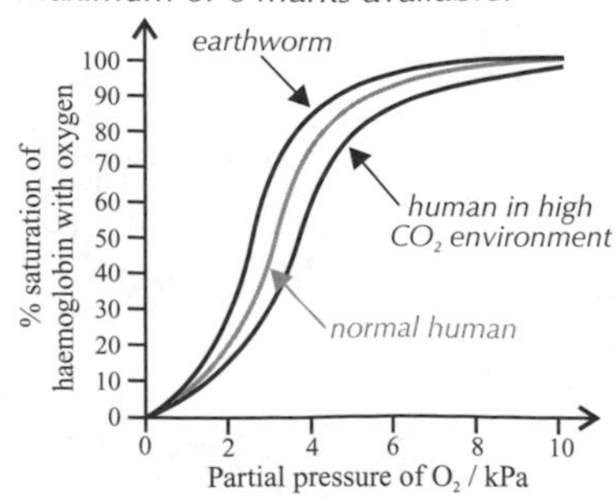


A curve to the left of the human one for the earthworm **[1 mark]**. Dissociation curves to the left indicate a higher affinity for oxygen at lower partial pressures **[1 mark]**. This enables the earthworms' haemoglobin to be saturated at the lower partial pressures underground **[1 mark]**.
A normal human dissociation curve that has shifted down for the human in a high carbon dioxide environment **[1 mark]**. This is the Bohr effect **[1 mark]**. High concentrations of carbon dioxide increase the rate of oxygen unloading and the saturation of blood with oxygen is lower for a given $pO_2$ **[1 mark]**.

2 It is composed of more than one polypeptide chain **[1 mark]**.
The reason that haemoglobin has a quaternary structure is because it has more than one polypeptide chain. The fact that it's made up of four polypeptides isn't important.

### Page 61 — Variation in Carbohydrates and Cell Structure

1 Maximum of 6 marks available, from any of the 8 points below. Starch is made of two polysaccharides of alpha-glucose **[1 mark]**. Amylose is a long unbranched chain **[1 mark]** which forms a coiled shape **[1 mark]**. This coiled shape is very compact, making it good for storage **[1 mark]**. Amylopectin is a long, branched chain **[1 mark]**. Its side branches make it good for storage as the enzymes that break it down can reach the glycosidic bonds easily **[1 mark]**. Starch is insoluble in water **[1 mark]**. This means it can be stored in large quantities without bloating the cells by osmosis **[1 mark]**.

## Unit 2: Section 4 — The Cell Cycle and Differentiation

### Page 63 — The Cell Cycle and DNA Replication

1 a) Maximum of 1 mark available.

**[1 mark]**

b) Maximum of 2 marks available. 1 mark for each new correct molecule with correct labels.

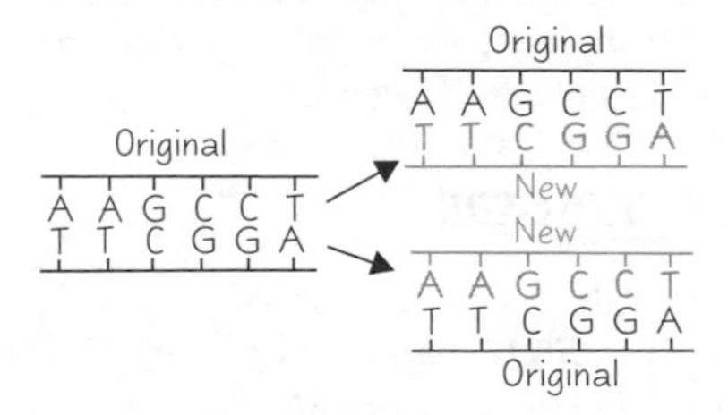


### Page 65 — Cell Division — Mitosis

1 a) Maximum of 6 marks available.
A = Metaphase **[1 mark]**, because the chromosomes are lined up at the middle of the cell **[1 mark]**.
B = Telophase **[1 mark]**, because there are now two nuclei and the cytoplasm is dividing to form two new cells **[1 mark]**.
C = Anaphase **[1 mark]**, because the centromeres have divided and the chromatids are moving to opposite ends of the cell **[1 mark]**.
If you've learned the diagrams of what happens at each stage of mitosis, this should be a breeze. That's why it'd be a total disaster if you lost three marks for forgetting to give reasons for your answers. Always read the question properly and do exactly what it tells you to do.

b) Maximum of 3 marks available:
X = Chromatid **[1 mark]**.
Y = Centromere **[1 mark]**.
Z = Spindle fibre **[1 mark]**.

# Answers

## Page 67 — Cell Differentiation and Organisation

1 *Maximum of 4 marks available.*
*The cell contains many microvilli/folds **[1 mark]** which increase the surface area for absorption **[1 mark]**. The cells form a layer just one cell thick **[1 mark]**, forming a short pathway for the nutrients to cross **[1 mark]**.*

2 *Maximum of 2 marks available.*
*It's best described as an organ **[1 mark]** as it is made of many tissues working together to perform a particular function **[1 mark]**.*

# Unit 2: Section 5 — Exchange and Transport Systems

## Page 69 — Size and Surface Area

1 *Maximum of 3 marks available.*
*Large mammals have a high demand for oxygen and glucose, which cannot be met by diffusion alone **[1 mark]**. This is because they have a small surface area:volume ratio **[1 mark]** and there is a large number of cells deep inside the body **[1 mark]**.*

## Page 71 — Gas Exchange

1 *Maximum of 6 marks available.*
*Gaseous exchange surfaces have a large surface area **[1 mark]** — e.g. mesophyll cells in a plant (or any other suitable example) **[1 mark]**. They are thin, which provides a short diffusion pathway **[1 mark]** — e.g. the walls of tracheoles in insects (or any other suitable example not already mentioned) **[1 mark]**. There is a steep diffusion gradient, which is constantly maintained **[1 mark]** — e.g. blood flowing in the opposite direction to water in fish gills (the counter-current system) (or any other suitable example not already mentioned) **[1 mark]**.*

2 *Maximum of 2 marks available.*
*Sunken stomata and hairs help to trap any water vapour that does evaporate **[1 mark]**, reducing the concentration gradient from leaf to air, which reduces water loss **[1 mark]**.*

## Page 73 — The Circulatory System

1 *Maximum of 6 marks available.*
*They have thick, muscular walls **[1 mark]** to cope with the high pressure produced by the heartbeat **[1 mark]**. They have elastic tissue in the walls **[1 mark]** so they can expand to cope with the high pressure produced by the heartbeat **[1 mark]**. The inner lining (endothelium) is folded **[1 mark]** so that the artery can expand when the heartbeat causes a surge of blood **[1 mark]**.*

2 *Maximum of 4 marks available.*
*At the start of the capillary bed, the pressure in the capillaries is greater than the pressure in the tissue fluid outside the capillaries **[1 mark]**. This means fluid from the blood is forced out of the capillaries **[1 mark]**. Fluid loss causes the water potential in the blood capillaries to become lower than that of the tissue fluid **[1 mark]**. So fluid moves back into the capillaries at the vein end of the capillary bed by osmosis **[1 mark]**.*

## Page 75 — Water Transport in Plants

1 *Maximum of 4 marks available.*
*In the apoplast pathway **[1 mark]**, water passes through cell walls **[1 mark]**. In the symplast pathway **[1 mark]**, water passes from cell to cell through the plasmodesmata that connect the cytoplasm of adjacent cells **[1 mark]**.*

2 *Maximum of 4 marks available.*
*Loss of water from the leaves, due to transpiration, pulls more water into the leaves from the xylem **[1 mark]**. There are cohesive forces between water molecules **[1 mark]**. These cause water to be pulled up the xylem **[1 mark]**. Removing leaves means no transpiration occurs, so no water is pulled up the xylem **[1 mark]**.*
It's pretty obvious (because there are 4 marks to get) that it's not enough just to say removing the leaves stops transpiration. You also need to explain why transpiration is so important in moving water through the xylem. It's always worth checking how many marks a question is worth — this gives you a clue about how much detail you need to include.

# Unit 2: Section 6 — Classification

## Page 77 — Principles of Classification

1 a) *Maximum of 2 marks available.*
*Green monkey **[1 mark]** because it's the closest to humans on the tree **[1 mark]**.*
b) *Maximum of 1 mark available.*
*Pig **[1 mark]**.*

2 *Maximum of 2 marks available.*
*A group of similar organisms **[1 mark]** able to reproduce to give fertile offspring **[1 mark]**.*
Make sure you use the word <u>fertile</u> in your answer.

3 *Maximum of 5 marks available. 1 mark for each correct answer.*

| Kingdom | Phylum | Class | Order | Family | Genus | Species |
|---|---|---|---|---|---|---|
| Animalia | Chordata | Mammalia | Primates | Hominidae | Homo | sapiens |

4 *Maximum of 3 marks available.*
*E.g. you can't study the reproductive behaviour of extinct species **[1 mark]**. Some species reproduce asexually **[1 mark]**. There are practical and ethical issues involved in studying some reproductive behaviour **[1 mark]**.*

## Page 79 — Classifying Species

1 *Maximum of 5 marks available.*
*DNA from two species is collected, separated into single strands and mixed together **[1 mark]**. Where the DNA is similar, hydrogen bonds will form between the base pairs **[1 mark]**. The more similar the DNA the more hydrogen bonds will form **[1 mark]**. The strands are heated and the temperature at which they separate is recorded **[1 mark]**. A higher temperature will be needed to separate DNA strands from more similar species because more hydrogen bonds will have formed **[1 mark]**.*
You need to use the right terminology here to get the marks, e.g. hydrogen bonds (not just bonds) and base pairs (not just bases).

2 a) *Maximum of 1 mark available.*
*Mouse and rat **[1 mark]**.*
b) *Maximum of 1 mark available.*
*Chicken **[1 mark]**.*

# Answers

## Unit 2: Section 7 — Evolving Antibiotic Resistance

### Page 81 — Antibiotic Action and Resistance

1 a) *Maximum of 3 marks available.*
*A mutation occurs in the DNA of a bacterium **[1 mark]**. If the mutation occurs in a gene it may alter the protein that gene codes for **[1 mark]**, which may make the bacteria resistant to an antibiotic **[1 mark]**.*

b) *Maximum of 3 marks available.*
*Resistance to antibiotics is spread between two bacteria by horizontal gene transmission **[1 mark]**. The two bacteria join together by a process called conjugation **[1 mark]** and a copy of a plasmid carrying a gene for antibiotic resistance is transferred from one cell to the other **[1 mark]**.*

c) *Maximum of 4 marks available.*
*Penicillin inhibits an enzyme involved in making the bacterial cell wall **[1 mark]**. This prevents cell wall formation in growing bacteria and weakens the wall **[1 mark]**. Water moves into the cell by osmosis **[1 mark]**. The weakened cell wall can't withstand the increased pressure so bursts (lyses), killing the bacterium **[1 mark]**.*

### Page 83 — Antibiotic Resistance

1 *Maximum of 6 marks available.*
*As the use of the pesticide increased, the number of aphids fell **[1 mark]** as they were being killed by the pesticide **[1 mark]**. Random mutations may have occurred in the aphid DNA, resulting in pesticide resistance **[1 mark]**. Any aphids resistant to the pesticide were more likely to survive and pass on their alleles **[1 mark]**. Over time, the number of aphids increased **[1 mark]** as those carrying pesticide-resistant alleles became more common **[1 mark]**.*

2 *Maximum of 3 marks available.*
*Some bats in the population will carry a mutation for a longer tongue **[1 mark]**. The bats with longer tongues will be able to feed from the flowers and so will be more likely to survive, reproduce and pass on their alleles **[1 mark]**. Over time, this feature will become common in the population **[1 mark]**.*

### Page 85 — Evaluating Resistance Data

1 *Maximum of 2 marks available.*
*Argument for: E.g. if a person has an infection that can be treated they should not be denied treatment. / If the person is not treated they may become very ill **[1 mark]**.*
*Argument against: E.g. a person suffering with dementia may forget to take their medication and so increase the risk of antibiotic resistant bacteria developing **[1 mark]**.*

2 a) *Maximum of 2 marks available.*
*E.g. the chest infection is mild **[1 mark]**. Prescribing antibiotics for non-life threatening illnesses contributes to increased antibiotic resistance **[1 mark]**.*

b) *Maximum of 2 marks available.*
*E.g. not prescribing antibiotics could reduce her son's quality of life **[1 mark]**. He may take longer to get better without antibiotics, so she may have to take longer off work **[1 mark]**.*

3 *Maximum of 3 marks available, from any of the 4 points below.*
*E.g. getting information from one area (East Anglia) would not show national trends **[1 mark]**. It's a relatively small study (only 300 patients), which decreases its reliability **[1 mark]**. Patients don't always tell the truth on questionnaires, which reduces its reliability **[1 mark]**. Patients may not return the questionnaire **[1 mark]**.*

## Unit 2: Section 8 — Species Diversity

### Page 87 — Human Impacts on Diversity

1 a) *Maximum of 2 marks available.*
*The number of different species **[1 mark]** and the abundance of each species in a community **[1 mark]**.*

b) *Maximum of 4 marks available.*
*Site 1 —*
*51 (51 – 1) = 2550*
*15 (15 – 1) + 12 (12 – 1) + 24 (24 – 1) = 894*
*Use of N (N – 1) ÷ Σn (n – 1) to calculate diversity index of 2550 ÷ 894 = 2.85*
***[2 marks for correct answer, 1 mark for incorrect answer but correct working]**.*
*Site 2 —*
*132 (132 – 1) = 17292*
*35 (35 – 1) + 25 (25 – 1) + 34 (34 – 1) + 12 (12 – 1) + 26 (26 – 1) = 3694*
*Use of N (N – 1) ÷ Σn (n – 1) to calculate diversity index of 17292 ÷ 3694 = 4.68*
***[2 marks for correct answer, 1 mark for incorrect answer but correct working]**.*
It's always best if you put your working — even if the answer isn't quite right you could get a mark for correct working.

c) *Maximum of 2 marks available.*
*The diversity of bumblebee species is greater at site 2 **[1 mark]**. This suggests that field margins increase the diversity of bumblebee species **[1 mark]**.*

### Page 89 — Interpreting Diversity Data

1 a) *Maximum of 2 marks available.*
*Both woodland and farmland populations have declined since 1970 **[1 mark]**. Farmland species have declined more than woodland species **[1 mark]**.*

b) *Maximum of 3 marks available, from any of the 5 points below. (or other suitable answers).*
*Loss of habitat **[1 mark]**.*
*Fewer hedgerows/larger fields **[1 mark]**.*
*Deforestation/clearance of land **[1 mark]**.*
*Farming intensification/changes to farming practice **[1 mark]**.*
*Pesticides causing disruption in the food chain **[1 mark]**.*

c) *Maximum of 8 marks available, from any of the 14 points below. Maximum of 4 marks available for benefits. Maximum of 4 marks available for risks.*
*Agriculture benefits — more food can be produced/there is an increased yield **[1 mark]**. Food is cheaper to produce, so prices are lower **[1 mark]**. Local areas become more developed by attracting businesses **[1 mark]**.*
*Agriculture risks — natural beauty is lost **[1 mark]**. Diversity is reduced because of monoculture **[1 mark]**. Diversity is reduced because of land and hedgerow clearance **[1 mark]**. Diversity is reduced from use of pesticides/herbicides **[1 mark]**.*
*Deforestation benefits — wood is provided as well as access to other resources **[1 mark]**. More land is available for homes/agriculture **[1 mark]**. Local areas become more developed by attracting businesses **[1 mark]**.*
*Deforestation risks — less carbon dioxide is stored, which contributes to climate change **[1 mark]**. Potential medical/scientific discoveries are lost **[1 mark]**. Natural beauty is lost **[1 mark]**. Diversity is reduced/extinctions may occur **[1 mark]**.*
When the question asks you to discuss an issue, you need to make sure you talk about both sides — the benefits and the risks.

# Index

# Index

# Index